Making Meaning with Machines

T0177296

Making Meaning with Machines

Somatic Strategies, Choreographic Technologies, and Notational
Abstractions through a Laban/Bartenieff Lens

Amy LaViers and Catherine Maguire

The MIT Press
Cambridge, Massachusetts
London, England

The MIT Press would like to thank the anonymous peer reviewers who provided comments on drafts of this book. The generous work of academic experts is essential for establishing the authority and quality of our publications. We acknowledge with gratitude the contributions of these otherwise uncredited readers.

This book was set in Stone Serif and Stone Sans by Westchester Publishing Services. Printed and bound in the United States of America.

Library of Congress Cataloging-in-Publication Data

Names: LaViers, Amy, author. | Maguire, Cat, author.
Title: Making meaning with machines : somatic strategies, choreographic
   technologies, and notational abstractions through a Laban/Bartenieff
   lens / Amy LaViers and Catherine Maguire.
Description: Cambridge, Massachusetts : The MIT Press, [2023] | Includes
   bibliographical references and index.
Identifiers: LCCN 2022052204 (print) | LCCN 2022052205 (ebook) |
   ISBN 9780262546126 | ISBN 9780262375139 (epub) | ISBN 9780262375122
   (adobe pdf)
Subjects: LCSH: Robots—Motion. | Human mechanics. | Movement, Aesthetics
   of. | Movement notation. | Bartenieff Fundamentals (Service mark) |
   Laban, Rudolf von, 1879–1958—Influence.
Classification: LCC TJ211.4 .L43 2023 (print) | LCC TJ211.4 (ebook) |
   DDC 670.42/72—dc23/eng/20230109
LC record available at https://lccn.loc.gov/2022052204
LC ebook record available at https://lccn.loc.gov/2022052205

10   9   8   7   6   5   4   3   2   1

To our teachers and our students, who are one and the same

Contents

Preface

When I was a PhD student in electrical engineering at Georgia Tech, I was given the most incredible opportunity: to study movement. I found myself surrounded by roboticists excited by the opportunity to learn something from a dancer. They seemed to say: Explain it to us. Help us make it. Now help us make it cool. Despite the many challenges of graduate school, I often found myself marveling at the immense privilege of getting paid to learn about my favorite subject.

First, I turned to my practice. The first paper core to my dissertation was inspired in a ballet class at the Atlanta Ballet School. The work developed an idea about smaller snippets of movement combining to form longer more complex phrases (LaViers & Egerstedt, 2011). One shortfall of the work is that it only specified a sequence of static poses, leaving the movements between poses to the imagination. So, the natural next step that I needed to address was: what happens between poses? Naturally, I turned to Rudolf Laban.

My teacher from about 1999–2005 (when I was in middle and high school), Irena Linn, had studied in Germany at a school set up by Mary Wigman (one of Laban's students). Miss Linn, as we called her, taught choreography through the use of taxonomy. She used her taxonomy to help us notice our habits, inspire new ideas, and create balance in our work. I remember the tiny chalkboard in the corner of Dancer's Studio in Knoxville, Tennessee. It was such an unusual feature for a dance studio, but she would have us crowd around it while she wrote the names of (and, I believe, symbols for) different movements on the board: rotation, jump, travel, and so on. She talked to us about changing levels and varying the dynamic quality of our movement (this one was always hard for me!). And

I still remember the first movement I choreographed in this new context: I pressed my hands, fingers spread wide, down on the floor next to my right foot; then, contracting from my core for support and keeping my foot and hands in the plane that had been formed by the smooth wooden planks on the floor, I lifted these distal parts through the air so that the sole of my right foot and the palms of my hands faced my audience. The movement displayed level change as well as forcing a weighty, strong physicality—and associated quality of motion—that did not come naturally to me.

So, it is likely that at some time in those early days, I heard Laban's name. I learned more about his work as I completed my senior thesis at Princeton University, where I took the novel opportunity to participate in a deep, independent, yearlong study to explore how the tools I'd been learning in my studies in mechanical and aerospace engineering (e.g., root locus analysis for understanding how controller gains affected closed-loop system dynamics) could inform the analysis that I was doing in my dance studies (e.g., comparing the styles of various choreographers working across genres and time periods). This is when I first encountered Labanotation and the effort system, feeling simultaneously excited by and dissatisfied with both: I marveled at the idea of a system of notation that could mark down the idea of a movement phrase just as music notation does for pianos, trumpets, and harps, but I wished for a system that would be used as regularly by dancers as music notation is by musicians. In those days, I felt so much optimism that quantitative tools could help create a new, more robust way of representing movement. My thesis was advised by Professor Naomi Leonard, who had so successfully wielded quantitative models in producing robot motion and capturing aspects of fish behaviors with collaborators outside engineering like Professor Iain Couzin. During this work, she introduced me to Professor Magnus Egerstedt, who would become my graduate adviser and whose motion-capture studio I borrowed to complete the work. In the end, I produced strange looping and wiggling plots that perhaps reflected stylistic differences between modern dance and ballet (but more likely just displayed the complex nonlinearity of human motion, a hint that Professor William Bialek gave me at the time, but that I did not understand until many years later).

Naomi's success in working with a biologist like Iain painted a tantalizing mirage about interdisciplinary research and gave me the idea that working with the field of dance as an engineer was possible. Perhaps one day this will be true, but I do not think it is today. There are a couple of important distinctions between biology and dance that must be highlighted. First, while biology and dance are both rooted in qualitative description (I am thinking of biologists venturing out of the lab into the natural environments of animals to *observe* their behavior and *write* their findings), there is centuries of work in laboratories using quantitative measurements of animal and other natural phenomena that sits comfortably alongside this observational work. Second, and maybe more important, there are lots and lots of books about biology. Of course, there are lots and lots of books about dance, too—there are lots and lots of books, period—but my sense is that there are more books, papers, and archived information about biology. To try to quantify this sense, the search term "fish behavior" currently returns almost four million results on Google Scholar, while "Laban Movement Analysis" returns about 40,000 and "Laban system" returns about 70,000, including a book called *Laban for All* by Jean Newlove and John Dalby (2004).

I opened that exact book in 2010, trying to understand the elusive idea of movement quality as I extended my initial research with Magnus. Newlove had been a student of Laban and worked with both dancers and actors (alongside Dalby) throughout her career. I consumed their book with excitement, and in the margins of my copy, you can see my excitement at quantifying this system. My first note adorns a description of "the whole-step," where I wrote, "An example of how [Laban] gave names to the elements of the cycle we see in joint-space" (Newlove & Dalby 2004, p. 20). In my recently completed undergraduate research, I had performed a linear (and nonlinear) decomposition of motion-capture data about walking, so this description refers to the cycles that appear on two-dimensional plots (projections of higher-dimensional spaces) in those types of analyses. After this description, they invite the reader to move, writing, "Try these time-honored step sequences but don't just do them with your feet. Let the movement flow through your whole body" (p. 20).

Years later, I am still untangling those words. Of course, it is impossible to take a step without moving your whole body, but Newlove and Dalby's point is clear: notice your whole body and make active choices within it. For my part, I was then, as I am now, obsessed with the idea of cataloging the incredible vastness of human motion. I think that at that time, even after decades spent in a dance studio, I had thought that the picture in my mind of a skeleton evolving in a high-dimensional state space (its joints plus their velocities) was complete, and this domain just needed a little math to sort itself into a satisfyingly simple picture. My thinking was something like: "Sure, the human body is a little more complex than a piano, but computers have lots of transistors and can help us find a basis for human motion that easily explains what we perceive in it." Today, I think that that line of thinking is laughably naive. Today, I think that it is only the richness of the space—and reveling in it—that can be satisfying. Then, guided by the limited texts I could find on Laban's effort system, including *Laban for All*, I happily modeled the quality of a movement with four continuous variables that figured into an optimal control problem, which was the extension that Magnus and I devised to include Laban and expressive quality into robotic motion (LaViers & Egerstedt, 2012).

Fast forward a few years, and I am an assistant professor in charge of my own research funds. My very first investment was going to be to study Laban's work for myself, at the source, in the studio. Professor Lori Teague, a member of the dance faculty at Emory University who collaborated on my graduate work in the effort system (LaViers et al., 2014), had connected me with a training program in movement analysis that I eagerly began. In the program, I was quickly immersed in a world that resisted publishing and regarded my beloved *Laban for All* with annoyance for its simplicity and audacity. Most of the books for the training program were primary sources: Laban and Irmgard Bartenieff's own writing, which was decades old by then. On the other hand, my graduate adviser had proudly displayed his multiple recently published texts that he had authored—which were in open, even joyous, competition with the texts written by his colleagues. He had taught me to see this kind of competition as a productive space for progress. These two worlds could not have been more different.

My time studying Laban's work as an engineering department faculty member led me to a very pessimistic outlook of engineering's capacity. The realities of getting funding, publishing, and advising students ruined the rosy picture of interdisciplinary research that Naomi had painted for me of her collaboration with Iain. Program managers insisted that I extend existing models claiming that Laban's effort system revealed the emotional content of movement (and I lacked a textbook to cite their mistake); reviewers judged my habit of coauthoring with movement experts as unethical (and I fought prior precedents where these collaborators were anonymous); my faculty colleagues told me that art wasn't engineering (leaving me confused about why I had been hired); and students were more interested in quantitative modeling than they were in qualitative reasoning (often at the expense of quality work). Railing against such simplistic, one-dimensional insistences, I tried to show roboticists how *inexpressive* their machines were (LaViers, 2019a), comparing their capacity to the staggering computing power of microchips and the marvel of natural life (LaViers, 2019c). I showed how important the arts were to robotic development (LaViers et al., 2018; Cuan et al., 2018; Ladenheim & LaViers, 2021) and how expressive *all* motion can be, experimentally demonstrating that emotive labels for movement break down across contexts (Heimerdinger & LaViers, 2019). The chasm between disciplines is fertile, if also turbulent.

It was in this context where I met my coauthor, Catherine Maguire, from whom I began taking classes—both within and outside my movement analysis certification program. Cat quickly introduced me to two books that were central to the process of complicating my relatively simple picture of human movement (a motion-capture body moving through a knowable state space). First, she had me complete coloring exercises from an anatomy book, a task that felt so *silly*, but which was responsible for teaching me, at the ripe age of twenty-six, facts as basic as the following: my stomach is *under* my ribs and my legs start *deep* inside my pelvis (I'm still working to find the extent of that depth today, a process that literally brings free-flowing tears to my eyes). She also gave me *Everybody Is a Body* by Karen Studd and Laura Cox (2013/2020), a book that I read with excitement but that left me aching for a more academic presentation of a system for movement analysis.

Where my knowledge is broad, Cat's is deep, and her forty years of experience in dance and movement studies create the depth of this book. Where I crave a sentence written with clarity, Cat craves a body moving with clarity (after all, her weekly class in Charlottesville, Virginia, is called "The Articulate Body"). Where I need a system with parallel elements, Cat wants to know what it *means* to an audience—especially her students. I do not know how to describe so much of what Cat has offered me: it is that kind of deep, nonverbal, maternal love that words fail to capture. Writing this book with her has been a true joy.

Cat and I began conceiving of this book at the end of 2019 and began writing in earnest during the peak of the COVID-19 pandemic. Our process consisted of biweekly, hours-long Zoom calls and lots of arguments. I remember one of the first arguments, which would continue over and over during our writing, about notation—specifically, the form of notation discussed in this book: motif. I often struggled to communicate the value and nature of movement analysis to an engineering audience—especially in funding applications, where space limitations frequently required brief, oversimplified, cartoonlike surveys of wide, broad fields. In such applications, I often described motif as something like a "shorthand" of Labanotation. This immediately connected motif to an idea that engineers could grasp ("movement notation") without entirely conflating it with Labanotation (a subtlety that was important to me). Cat rejected that term strongly, insisting motif was about "essence." Well, despite almost a decade of training with and working alongside her, I barely knew what she meant by that—and I certainly couldn't communicate it to a funding body or in a book proposal (and nor could she).

We returned to this argument often, and it shaped the book, which was not meant to be about movement notation (our initial plan was simply to have one chapter briefly dedicated to the mechanics of it), but we discovered a deeper connection between naming and notating movement ideas than even we had anticipated. As such, we have produced a book that culminates in a chapter on notation, building the entire time toward motif as a final goal, rather than a technical aside. Right now, I feel like the next book will go further into that forest and be entirely focused on the topic of notation. It is how we will solve the segmentation problem (LaViers & Egerstedt, 2014; Sheng & LaViers, 2014). It is how we will make safer autonomous vehicles, mend our relationships with our smartphones, and otherwise better incorporate machines into human experience. It is how we will talk to aliens.

And so, it was while standing on the edge of that forest that we heightened our goals for presenting the symbols used in the book. I first met Jonathan Pearce working on an online automation design tool, where I learned of his ability to weather robust argument to clarify ideas. Asking him to help create some of our more complex illustrations, as well as design a consistent symbol set for motif, was a wonderful decision. Jon's artistry and expertise in graphic design have helped Cat and me sift through many iterations of the visual cues and conventions that each symbol should be composed of, in order to best relate to other existing systems as well as prior renderings of that symbol—a process which has interestingly revealed many oversights in our thinking.

As we go to press, we have found a funding source to enable the open-access publication of this book, part of a National Science Foundation (NSF) infrastructure grant that will create a shared resource of video clips of human motion labeled by expert annotators. The grant, a collaborative effort between Penn State University, University of Illinois Chicago, and my lab, is aimed at facilitating the creation of new tools in computer vision, human-robot interaction (HRI), and the study of human movement more broadly. With each iteration of developing the proposal, we have adjusted it to grapple with the personal nature of meaning and to dive deeper into the nuance of the annotation process. I remain hopeful that in this team, we have a match like Naomi and Iain, and I am grateful to be bringing a true textbook on movement studies into the collaboration—and to share with a community of engineers interested in the expressive dimensions of human motion more broadly.

Researchers rely on the richness of their internal movement model to inform the questions they ask and the answers they are willing to accept. In my doctoral work, I thought that four continuous variables might capture some of the distinct qualities of human movement. Since then, I have suggested that the 3,240,000 static, discrete poses that can be measured by a typical motion-capture recording of a human body are not enough to encompass the incredible phenomenon of our bodies in motion. Today, I feel that even this book barely scratches the surface in cataloging human movement; but somewhere out there, I know that a dancer is being asked by an engineer to explain herself; and I hope this book moves her answer farther than mine and deepens her own questions.

Amy LaViers

Philadelphia, Pennsylvania
June 30, 2022

Acknowledgments

Writing this book is the result of years of work between us and our communities. We want to give special thanks to several members of those communities in which our understanding of movement studies was challenged and grew as a result. First and foremost, we want to thank the Robotics, Automation, and Dance (RAD) Lab members and collaborators, including James Wang, Rachelle Tsachor, Tal Shafir, Karen Bradley, Kate Ladenheim, Reika McNish, Wali Rizvi, Ilya Vidrin, Riley Watts, John Toenjes, Maxwell Asselmeier, Erin Berl, Alexandra Nilles, Allison Bushman, Jacey Lambert, Jordan Parker, Umer Huzaifa, Joshua Schultz, Caleb Fuller, Catie Cuan, Roshni Kaushik, Jighisha Sampat, Omar Darwish, Yichen Zhou, Ishaan Pakrasi, Alexandra Bacula, Varun Jain, Novoneel Chakraborty, Joel Meyer, Madison Heimerdinger, Alexander Zurawski, Jamie Sergey, Lin Bai, Jialu Li, Anum Jang Sher, Hang Cui, Manav Sanghvi, Kim Brooks Mata, Zachary Calhoun, Colleen Kohout, Brett Libowitz, Anne Moenning, Crispin Bernier, Benton Turnbull, Jason Ye, Jerry Heddy, Ashkan Bashiri, and Yu Sheng. We are also grateful to our collaborators in the Certified Movement Analyst (CMA) community, especially the coterie of WholeMovement, Laura Cox, Esther Geiger, and Karen Studd. We especially thank those who participated in interdisciplinary workshops in the RAD Lab alongside us. Their intellect, embodiment, and approach to movement studies have significantly affected our own.

We are grateful to the National Science Foundation (NSF) for funding the open-access of the book through award 2234195. We want to acknowledge funding from the US government that supported the research described throughout the book, including award D16AP00001 from the Defense Advanced Research Projects Agency (DARPA) and awards 1528036 and

1701295 from NSF. We also acknowledge the physical spaces where this work developed, including the dance and mechanical science and engineering departments at the University of Illinois Urbana-Champaign; the systems and information engineering department at the University of Virginia; the McGuffey Art Center in Charlottesville, Virginia; AgapeBelgium in Ardooie, Belgium; and the Conference for Research on Choreographic Interfaces at Brown University (where we had our inaugural writing session).

We thank our colleague Elizabeth Jochum for putting us in touch with Doug Sery at the MIT Press, where we were so amazed to have found a simultaneously sympathetic and critical voice. Moreover, our thanks go to our current editor, Noah Springer, whose guidance has greatly helped our work. Both Doug and Noah recruited amazingly thoughtful reviewers who lent significant time and intellect to reviewing the first early chapters, and then an entire manuscript. The book has shifted and changed with each round of their feedback, for which we are incredibly grateful. Jonathan Pearce's wonderful symbol designs are featured throughout the book; we are so grateful to his talent and energy in working with us on these illustrations (and allowing us 1.1 pt linewidth; that extra 0.1 is everything!). Our colleague Penny Chang reviewed an early draft and has been very helpful in creating clarity in our thinking and writing, and Kate Ladenheim reviewed a final draft, giving us important feedback from someone teaching and working right at the intersection of technology and dance. We also owe a great thank-you to Nicholas Osborne as well as the production teams at MITP and Westchester Publishing Services, who proofread, formatted, and copyedited this manuscript with a detailed, discerning eye. Finally, we thank the very first readers of our work, family members who also supported us through an intense experience: David Campanelli, Keegan Campanelli, Bryce Campanelli, Margaret LaViers, and Eric Minnick.

Electronic Resources

Electronic resources are available at www.makingmeaningwithmachines .com.

Materials for the **symbol set** presented in chapters 5–8 and appendix A include digital renderings and editable files of each symbol, as well as additional symbols.

Materials for the **notation** presented in chapter 10 include digital renderings and editable files of the scores in key figures, as well as video footage of the authors performing the notated movement phrases.

Materials for the **embodied exercises** introduced in chapter 3 and used throughout chapters 5–8 and 10 include video footage of the authors demonstrating several key exercises.

These materials are hosted online by the Robotics, Automation, and Dance (RAD) Lab.

Prelude: Opening with Embodied Perspectives

A butterfly floats erratically past my view. It seems to flit around chaotically, perhaps being eddied around by the slight breeze that moves the leaves of a nearby tree. Yet as I look more closely, its two large, wide wings flap down with clear control and intent, breaking the chaos with order and organized movement. A car approaches, slowing as it comes to a crosswalk. The path of the car follows the outline of the one-way street, and its linear path is a stark contrast to the butterfly: stable, orderly, bleak, clear. The car cuts through a large, brick pedestrian promenade at the heart of downtown Charlottesville, Virginia, where I sit on a shaded bench. Trees dot the space in irregular intervals, causing pedestrians to weave and wind as they are forced to make choices on how to accommodate the irregularly spaced obstacles. It's Wednesday around lunchtime, and I can see clusters of people heading in and out of the various restaurants along the sides. This is an undeniably pleasant and restful moment. Looking inward, I stretch out my legs, feel my knee crack and pop, a slight sensation of burning, a sore muscle surrounding an old injury; moving will be both annoying and replenishing today. Standing, I begin to walk to the nearby McGuffey Art Center.

During my walk, I notice my experience of movement. There is a fuzzy line between what is and what I perceive: I don't notice the reason for a sudden tweak in my ankle until I look back and see the substantial rock I accommodated without conscious awareness. And yet I perceive so much. So many details. The sore hamstring that groans as I overtake a hill. The pinching in my left toes as I hurry to make class on time. The rhythm between my scapula and hands as they swing back and forth, in and out of pace with my feet. Arguably, these details comprise more than the handful of commands I design for robots, like the NAO humanoid with its fourteen movable degrees of freedom, where a known, countable set of motors creates the opportunity for actable movement. I'm here to take a dance class, in which a plethora of details from my own experience must be whittled down and simplified to help me choreograph motion—both for my own body and the artificial bodies I design.

Amy

As I enter my studio at McGuffey Art Center, I am struck by the golden sunlight streaming in through the windows and warming the floor. Cleaning the floor for my class, I push the mop along the parallel lines of hardwood planks, delineating the space. I feel my feet settling into the floor against the pressure of the mop and notice the changes in temperature of the floor surface as I move in and out of the areas warmed by the sunlight. I use this set of spatial pathways and my own body weight to activate a sense of my own agency in interacting with and changing the space where I am about to teach.

Today's lesson is on resiliency and the body's physical "core." I will explain how the spine is the key to adapt the body to new movement tasks and recover balance from a fall. To teach this concept, I will use movement sequences that activate an awareness of the vertebrae, illustrating how the spine is critically, although sometimes subtly, involved in every movement. Vertebrae can be sensed through a myriad of methods, such as direct palpation with the hand, descriptive imagery of our anatomy, and the weight shift that occurs during contact with the floor. Soon, I will ask my students to stand and begin to engage their sense of self by bouncing, jiggling, and breathing in order to better feel the viscera surrounding their spines and to understand how this relationship between stable bony elements and mobile soft tissue evolves in movement. Through this kind of attuning to inner sensation, the class will shift from inner awareness to outer awareness as the movement becomes more complex: from the floor to standing, from standing to traveling through space, from traveling through space to partnering with another dancer, from partnering back to stillness. The class will progress from an inner sense of self to a relationship to the outer environment, relating sensation of internal body parts to actionable changes in the environment. The goal of this work is to offer my students opportunities to make choices in how they engage with their environment and relate their own moving containers to the larger world.

Cat

Introduction: Inviting Engineers into Movement Studies

This book was written by two dance artists: the student in the prelude (Amy), employed as a roboticist, and the teacher in the prelude (Cat), focusing on inner awareness of physical expression in her approach to movement instruction. This text results from the application of our shared training to the development of robotic and automated systems. The biggest shared component of that training is a particular, comprehensive system for analyzing movement that this book is grounded in, but we also bring broader training in dance, artmaking, engineering, and research.

On our journey to understand the phenomenon of human movement, we have established a rich collaboration spanning disciplines that has left us with more questions than concrete answers. How should a gestural interface react to a "flick" versus a "dab" versus a "punch"? Should robots reach out to a human counterpart with a direct, telescoping action or through a circuitous arc in space? How many distinct actions does a person perceive in a given robotic behavior? How do we know whether (and how) to yield to a colleague walking in the hallway? How will we know whether (and how) to yield to a twelve-armed robot "walking" in the hallway?

Patterns in movement seem to advertise internal states of human movers, and engineered systems increasingly aim to reflect and accommodate this idea. To help achieve this goal, this book is a primer for seeing, describing, and creating a wide array of movement patterns with both natural and artificial bodies, facilitating broader and better design choices for roboticists, technologists, designers, and artists. The main topics of the book are a descriptive and symbolic taxonomy for postural and gestural bodily movement and a theory for understanding how people create and perceive meaningful patterns in such movement. The book does not offer many equations, programs, or algorithms (although it cites many texts that do), but its material is grounded in examples and terminology that are relevant to roboticists, computer scientists, and technology designers.

Delving into Somatics, Choreography, Notation, and Machines

Among formalized systems of somatic practice, choreographic design, and notation is a broad suite of body-based theories and practices—a body of knowledge that we broadly term "movement studies," which includes movement analysis. Whereas most domains in this field focus on either inner experience (e.g., Body-Mind Centering/BMC) or outer expression (e.g., ballet technique), Laban/Bartenieff Movement Studies (LBMS)—a contemporary term meant to describe both Laban Movement Analysis (LMA) and Bartenieff Fundamentals (BF), along with advancements since the passing of the work's founders and namesakes Rudolf Laban and Irmgard Bartenieff—sits at the nexus of interior, body-based experience, where the somatic practice of movers is at the forefront of movement investigation, and the exterior, performative pursuit, where choreographic principles for creating observable changes in bodily motion serve the goal of presenting ideas on stage. Thus, this field of movement studies has implications not only for how we might design the movement of machines, but also why and how human movers experience that motion. In this book, we are not simply interested in cataloging this body of knowledge, but in exploring how this work has implications for designers, researchers, and technologists—namely, how new use cases for these relatively old tools will affect everyday life. That is, what will people think of robots in new, emerging contexts? Thus, we invoke more active language, and throughout this book, we refer to the *application* of somatic practices and choreographic principles in technology design as the use of somatic strategies and choreographic technologies, respectively.

As a new assistant professor, building a research program in robotics at an engineering school, Amy sought a way to train her growing research group, the Robotics, Automation, and Dance (RAD) Lab, which was filled with science and engineering students, as well as designers, dancers, and other artists, who had a variety of movement backgrounds (e.g., soccer, ballet, tai chi, playing the violin, and making bread) but had never used these experiences (at least formally) in technology design. To provide students with the tools to better reflect on and subsequently understand those experiences, Cat was invited to create workshops for these students, which translated into involvement in the students' research, growing this teaching relationship into a research collaboration that has investigated and begun to illuminate how first-person, embodied movement is a resource for work in robotics, artificial intelligence, and other technologies.

The workshops were held in dance studios (including the RAD Lab itself, as shown in figure 0.1) and felt more like dance classes than engineering courses (although they were both). Over the years, they were taught by both Cat and Amy, as well as other collaborators in dance and movement studies. The primary participants in the workshops were graduate students conducting research in robotics while enrolled in degree programs in engineering, but robotics is an inherently interdisciplinary field, so the workshops were also attended by artists, computer scientists, dancers, designers, kinesiologists, movement analysts, musicians, and somatic practitioners from the RAD Lab, as well as the broader university communities in which the workshops were held. These workshops were supplementary to typical activities in research on robotics, human-robot interaction (HRI), and engineering, giving students and employees in the RAD Lab experiences that were fully centered around the space of movement studies. These workshops are the basis for this book, and their attendees and content are representative of the audience and topics of the book, described in the next two sections.

In the remainder of this section, we will describe how we began studying gait with a doctoral student named Umer Huzaifa (now an assistant

Figure 0.1
Snapshots from the RAD Lab at the University of Virginia, taken from fall 2014 to spring 2015. Left and center-left: Cat teaching movement workshops to graduate and undergraduate engineering students. Center-right and right: Amy conducting research with the team, leveraging choreographic and somatic learnings from these workshops in the development of expressive robots.

professor at DePaul University in Chicago), when one of these workshops saved Umer's fledgling passion project, setting it on the path to becoming a core research area in the lab that would eventually lead to work published in top robotics venues, including the International Journal of Social Robotics (SORO), the International Conference of Intelligent Robots and Systems (IROS), and the IEEE International Conference on Biomedical Robotics and Biomechatronics (BioRob), and be funded by a grant from one of the most prestigious and competitive sources of research funding in the US: the National Science Foundation (NSF).[1] We will revisit Umer's work later in the book, but it is instructive to share here how it began in order to paint vivid details of the materiality of this kind of interdisciplinary work and, therefore, why we wanted to write a book to capture our workshops.

Umer was a graduate student in robotics with an undergraduate degree in electrical engineering, and he wanted to study gait for bipedal robots. He had read many papers on gait and bipedal walking robots; Russ Tedrake's work was a particular favorite. Amy had never worked on gait, but her training in dance and movement studies described the role of the core in human gait, which she knew was not emphasized in biomechanics or leveraged in robotic walking. This disparity provided a potential opportunity for new work. So, promising that he could work on this project in parallel to other work he was doing in the lab, she gave Umer a book of movement exercises and asked him to try them himself at home. This is not how somatic and choreographic thinking best develops (and yet it is the setting for you, our reader, as well).

After a few meetings, it was clear that there could be some interesting explorations of this topic. Amy found a large group of students willing to come in on the weekends and work on this special side project. The first meeting was a workshop led by Cat in the RAD Lab at the University of Virginia. When Cat entered the lab to work with the students, a division was immediately felt: she was dressed in dance clothes (form-fitting tights and bare feet), while the students were wearing street clothes (shorts, jeans, belts, socks, and shoes). The divide would go beyond fashion: each group valued different epistemological systems.

The first thing that Cat did was to ask the students to take off their shoes and socks and walk around the room, feeling their feet and their shifting weight as they traveled around. An internal sensing of the changes in pelvis as their weight shifted over their bare feet was encouraged with cues around feeling the bones, the rotational relationship of the leg and pelvis, and holding their hands on their hips to feel the movement through their own hands as it occurred. The next step was lying on the floor and introducing movements that would further support awareness of the pelvis, legs, and the interactions between them. To get a sense of the scene, imagine a group of students lying on the cold, linoleum-clad concrete floor of the lab, engaging in subtle movements designed to connect them to the internal space of their moving bodies.

While there was definitely some awkwardness, these students embraced the experience and were willing to do more in this vein of body research. Umer lingered at the end of this workshop to share his ideas with Cat about his research on gait and how this embodied experience supported his creative design process. This began the collaborative research relationship between Cat and the RAD Lab. Now, using Umer's work as a concrete, connecting example, we would like to illustrate how knowledge from somatics, choreography, and notation affected our work on gait design for robots.

First, how *somatics* has influenced our work with machines: Instead of approaching the parameterization of gait from external measurements, where ankle, knee, and hip displacement is the most salient feature of the movement, we considered an internal, somatic experience of gait (Huzaifa et al., 2016), as parameterized by concepts from Bartenieff Fundamentals, in particular "lateral pelvic shift," "sagittal pelvic shift," and "thigh lift," which describe the action of the pelvis side to side and front to back and the flexion at the hip joints, respectively. These exercises are typically used to retrain patients in physical therapy or students in a dance class, enlivening their pelvic core and enriching their access to weight shift from it. Thus, our parameterization focused on these three concepts—notably ignoring the action of the most distal joints in favor of proximal and core-based actions. This shift led to innovations in hardware design that have been applauded for their ability to mimic the natural dynamics of human walking, which is notably energy efficient: consider the difference between shifting the center of mass of a system directly, allowing gravity to power a fall (our approach), versus wrenching that center of mass through space via distant, distal actuators (the traditional approaches).

We were interested in developing a system with many, many styles of gait. To do this, we turned to *choreography*. Rather than working to create a single, energy-efficient gait, or even to categorize running versus walking (gaits distinguished by differences in footfall ground contacts) as many researchers do, we wanted to design many varieties, with subtle variations among them. We settled on searching the space of walking (gait defined by continuous contact with one of two bipedal lower appendages and the ground), such as skittering, sauntering, ambling, and dragging. These descriptors come from synonyms in the English language for "walking" (of which there are over sixty). We ended up validating six distinct walking gait styles in studies with human subjects, broadening the palette of options for machine movement designers (Huzaifa et al., 2020). None of these gaits use minimal energy (the mere presence of six valid gaits hints that we were not minimizing energy usage), but they offer a way of thinking about how distinct internal states may be communicated—or context may be reflected—through external, observable movement styles that are salient for human viewers.

The movement of the pelvis is almost imperceptible to most sensors that quantitatively measure and digitally map movement: one team of prior researchers had to surgically implant reflective markers to document its motion using infrared cameras (Crosbie et al., 1997). Thus, instead of by external measurement alone, it was through direct engagement with movement analysis that we leveraged somatic strategy and choreographic technology in robotic design. Or, as we wrote with a larger team of dance experts and engineering researchers:

> Our approach is pragmatic: we want to understand the phenomenon of how people create such vastly varied motion profiles that communicate complex intent. This knowledge, we believe, is contained inside body-based movement training and somatic practice where practitioners hone their own movement capabilities by expanding their array of choices. External methods, typically employed in the sciences and in engineering such as motion capture, photography, force plates, and the like, can work to document the result of a movement pattern but do not have access to choices made by a human in focus, motivation, sensation, memory, prior muscle patterning (and re-patterning), etc. The practice of honing these choices is one of embodiment—a body of knowledge that cannot be known but only moved. (LaViers et al., 2018, p. 3)

Unlike the computational simulation of gait, where we determined the pattern of forces necessary to produce forward locomotion in various styles on a particular, simple model of a given mass, the human labeling of these gaits is not universal across subjects and will change over time. In 1,000 years, the forces used to generate these gaits will measure the same, but culture, lifestyle, and conventions will have changed dramatically, creating new contexts for observing, naming, and interpreting the very same patterns in movement. If the pace of life increases over this time, what is today a hurried gait that transmits urgency and expediency (or means "I am in a hurry") may seem normative or neutral (or means "I am not in a hurry") in this distant future. Indeed, this future world is sure to have new styles of walking altogether. To grapple with the complexity of meaning-making, the book will often return to and emphasize the idea that meaning is specific to a given context (and therefore is not always shared between the mover and observer).

In addition, the representation of the gaits we designed affects how they are understood. Umer's simulation used vectors discretized in time to represent these gaits. Among the research team, we tried the gaits with our own bodies, moving them physically; we riffed on each one, varying it a bit, then a bit more, and so on, in order to come up with new varieties that we tried to name, notice, and codify; we displayed these gaits to subjects as rough cartoons; we came up with representative labels in English; and, in the future, we could use a symbolic system of *notation* to represent the gaits. Broadly, we can think of these alternative methods of representation as various abstractions—notational abstractions—that influence what is observed, experienced, interpreted, and saved for translation to other bodies. The interdependence of mover, context, and observer becomes especially poignant when notating movement. Notating movement is personal and embodied, and a form of meaning-making that can help clarify our design goals for machines.

A Growing Community at the Intersection of Robotics and Dance (Whom This Book Is For)

Researchers with significant formal training in somatics, choreography, and notation are rare in traditional engineering programs, and it is hard to study this material through books and archival publications. Most movement practitioners come from dance, therapeutics, and somatic practice, so the medium of their work is not primarily academic writing; rather, knowledge is passed through physical, bodily practice in studios, therapy rooms, and gyms. Nevertheless, the use of movement studies, especially in the Laban/Bartenieff tradition, is growing in engineering and computer science research.

A search conducted on the term "Laban" in June 2020 showed 383 papers in the Association for Computing Machinery (ACM) database and 105 papers in the Institute of Electrical and Electronics Engineers (IEEE) database, with 268 (70 percent) and 80 (76 percent) of these results published in the prior ten years. Searching "Bartenieff" yielded 23 papers at ACM and 8 papers at IEEE, with 15 (65 percent) and 6^2 (75 percent) published since 2012. Relatedly, a published data set using the query "dancer" in the ACM database returned 194 papers since 2019 (Rajko, 2021). While these papers represent a very small portion of the overall publication activity in these large professional societies, they represent a segment that is growing—and will grow much more in the coming years as robots, computers, and machine interfaces increase in daily human experience.

And so, in writing this book, we are targeting roboticists, technologists, designers, and artists who seek to utilize technology as a means of expressing and/or recognizing human movement, promoting the widespread idea that a deeper understanding of human movement will help us build better tools in an increasingly technology-based world. We imagine our primary audience as graduate students, technical researchers, and product designers who are working in areas of human-robot, human-computer, and human-machine interaction and need better background knowledge about choreography, somatics, and notation. In parallel, dancers, physical therapists, and movement coaches may use this book to better understand the potential of their expertise in technology development and of the use of technology in their creative process. Further, other movement analysts may enjoy this book as a reference for themselves, as well as a road map to the impact of this work in technology. Increasingly, this hybrid field drawing on movement studies (including dance, theater, and digital media) and robotics (including human factors, control theory, and mechanism design) is referred to as "expressive robotics," or even "choreobotics."

University courses in human factors, arts and engineering, HRI, affective computing, animation, human subject research, and other increasingly offered programming at the intersection of design, the arts, and engineering, may use this book as a primary reference. For example, Amy's graduate-level course in mechanical engineering "High-Level Movement Representation and Robotic Control,"[3] taught in 2017 and 2019 at the University of Illinois at Urbana-Champaign, is the kind of interdisciplinary course that could use this book to supplement traditional engineering texts with vital knowledge from movement studies. Similar courses are increasingly offered in higher education. They often feature faculty and lecturers from both engineering and performing arts, some of whom—especially those working in performance studies, dance, and kinesiology—may have specific training in choreography, somatics, and notation. Movement analysis certification programs (like the ones that Amy and Cat participated in) tend to bridge all three of these domains and also may find this text useful in demonstrating application areas for the work, as well as a reasonable reference source featuring symbols and their corresponding concepts identified in one place. In addition, traditional courses in HRI, design, and human factors may increasingly find this textbook useful as these disciplines continue to be enriched and broadened through interdisciplinary exchange, evidenced by the growing body of work that employs movement studies in technology development.

Industrial technology designers are increasingly thinking about embodied interactions between humans and artificial agents. Roboticists are working on the problem of "social robots," such as the Amazon Astro, that can seamlessly integrate into human-facing social situations. Software designers are using new off-the-shelf tools like Microsoft's Azure Kinect depth camera and software suite to create immersive experiences that react to users' bodily actions and gestures. Engineering programs are widening the scope of their course offerings to include physical, human-centered design to better prepare students to create devices for the Internet of Things. Yet how should such experts think about the body and its patterns in motion? It is not practical for every technologist to have extensive training in body-based design taxonomies and theory as well, and so our goal has been to write a book that makes this body of knowledge, accrued through physically taxing hours spent in a studio setting, more accessible to this community.

Finally, machines are already personal. As our tools increase in number and complexity, it is worth spending time grappling with how these devices change the narratives of and experiences in our daily lives. So, while this book is primarily written for professionals working at the intersection of the body and machines, it is also for anyone interested in interrogating their own relationship with machines. After reading this book, users who interact with a device such as a Fitbit may be better able to interpret the measures provided by that device inside their own personal contexts and bodies.

Meaning-Making with Machines (What This Book Is About)

In this book, we aim to create a broad reference on movement studies (theory and practice) appropriate for engineers, designers, and researchers designing new tools that either produce or interpret movement styles and patterns. Primarily, we identify LBMS as a useful framework for organizing, producing, and explicating somatic strategies, choreographic technologies, and notational abstractions. We are presenting an edited and expanded version of the system stemming from our ever-evolving work with machines, adding new and refined terminology, analysis, and notation.

We even suggest that working with machines is essential to developing a systematic understanding of perceived movement, drawing a parallel to how the piano and other instruments were essential to figuring out perceived patterns in sound. Thanks to comprehensive music notation, we have access to playing the same song on different instruments, algorithmic readability of music, music copyrights, and many more technologies that rely on this systematic, abstract understanding. We are not there yet with movement and dance, but such understanding is growing with the explication and expansion of the tacit, embodied research that goes on in dance studios, physical therapy rooms, gyms, and outdoor fields, where the passage of knowledge is person-to-person and body-to-body, rather than from person-to-book or book-to-person. This book aims to add another flow to that exchange, and while holding fast to the tenet that *not everything can be written down*, we provide notational abstractions to support the process of writing down movement.

This book tries to scale the experience of in-person events that we have conducted for interdisciplinary audiences, like the workshop that kicked off our work with gait, to an introductory text for researchers, developers, designers, and artists interested in understanding principles of somatics and choreography (what we term more broadly as "movement studies") and their application to the design of new systems and interfaces. In our workshops, we expose graduate and undergraduate students to different aspects of choreography, improvisation, somatics, meditation, notation, and observation through the lens of movement studies as tools for meaning-making through recognizing patterns, contexting, and gaining insight into the complexity of human movement. We approach these ideas through experiences that frame the body's movement as the basis for our understanding of the world. In measuring, analyzing, and noticing bodily movement, it is critical to consider both *qualitative* and *quantitative* analyses from *subjective* and *objective* perspectives. As such, this book will not feel like a typical text on developing novel robots. Furthermore, we do not imagine that one could develop a novel robot with only this book as a guide. It is instead meant to supplement the many existing works on computational programming, mathematical modeling, and empirical validation of machines.

For the purpose of sharing physical experiences through words on a page, the book uses "embodied exercises," where we make the somewhat unusual request that readers get up and move in their own environment. Often, these exercises will be supported through observation, writing, video recording (of yourself and others) and rewatching, and engaging other moving bodies. We know that many readers will not have access to the requisite tools, space, and time to complete such exercises in every reading of the book, so while we implore the dedicated readers to make time and space to experiment with these exercises, we have worked to create stand-alone text that explicates the physical, embodied knowledge that these exercises are aimed at transmitting. These exercises are the native format of the knowledge that this book presents, and being able to learn in this format will require practice. Perform the exercises, come back after some time, and try them again; film yourself and watch the playback; take notes; and use your body as a place for observation. Over time, you will perceive more from these exercises and thus be able to experience them with more nuance—a cyclic pair of reinforcing processes. As such, these exercises will not immediately impart physical wisdom, but trying them is a necessary step. Physical experiences will always be distinct from written text or derived equations, so we implore readers to take on these exercises with rigor and energy.

Chapter Road Map

Chapter 1 reviews prior literature that grapples with how to understand humans' embodied experience and expression. We review the philosophy of meaning-making, as well as methods of measuring and describing movement. We draw on a broad set of fields ranging from philosophy, mathematics, the social sciences, physics, and biology, which we cannot fully cover in depth. Instead, our aim is to share important references that have shaped our thinking, with an emphasis on scholars who are considering an internal, first-person point of view, as they are likely to be less familiar to our primary audience of technology designers. We expect these readers to be more familiar with concepts traditionally presented in robotics textbooks, so such material is less developed in this book. Moreover, the concepts in LBMS have historically been underpinned more directly by this body of thinking than by technical designs for machines. This chapter is meant to broadly inform and point readers to deeper references on these complex topics.

Chapter 2 zooms in on movement studies, the particular body of knowledge that we employ here, broadly describing types of somatic practice, choreographic process, and notation. Chapter 3 then focuses most directly on the book's intellectual center, LBMS, presenting a vocabulary of terms dealing with the creative generation of movement design (the process more generally identified as "choreography") and the interpretation of movement experience (the practice more generally identified as "somatics"). That chapter ends with the introduction of embodied exercises, which provide an opportunity for crucial investigation outside the pages of this book.

Thus, part I begins broadly, reviewing the literature on meaning-making, and ends with the specific goal of this book: the introduction of a taxonomy that drives a notation scheme for describing movement. The organization of the next five chapters, forming part II, is created by five components used as lenses to analyze movement, including a component that has emerged from our work with machines: Body, Space, Time, Shape, and Effort. In each associated chapter, we look broadly, congealing around questions often used to describe these categories and bringing our broader experiences in dance and dance-making to bear on these topics. These chapters form the bulk of the reference material of this book, and each contains a section describing the application of that chapter's content to machines.

In chapter 4, we introduce the *Body* component and ask "*What* is moving?" answering with the role of the form or container of a moving body and the somatic, internal experience of movement. In chapter 5, we introduce the *Space* component and ask "*Where* is the movement?" articulating broad spatial categories that help organize the happenings of a moving body and begin to relate this action to the environment. In chapter 6, we introduce the *Time* component and ask "*When* is the movement happening?" explaining how rhythm and phrasing help human movers come to fairly precise notions about timing of movement. In chapter 7, we introduce the *Shape* component and ask "*For whom* is the movement?" delineating the types of connection—with self, with another, or with a group—that help name the spatial patterns of movement in its environment. In chapter 8, we introduce the *Effort* component and ask "*How* is the movement executed?" providing options for the texture and tonality of movement that help name the temporal pattern of movement in its environment. This chapter outlines and extends Rudolf Laban's famous Effort System, which explicates categories of motion quality. (There are already numerous endeavors in the emerging field of expressive robotics that have tried to incorporate this system in machine design, including our own attempts, but we don't think that any of these works are mature enough yet to be called a complete success.) It also introduces the "affinities" between each of the components of the system, which interact to bolster the richness of human motion.

Any coherent movement phenomenon of a body provides answers to these five questions (what, where, when, for whom, and how) and indeed many others, expressing something meaningful to an intelligent observer. The categories that comprise part II, therefore, offer a limited picture of movement that needs to be resolved with synthesis and integration inside of applications. To that end, part III of the book makes the work specific to applications in human-machine interaction and design. In particular, it introduces a guiding framework for seeing and notating bodily motion, with an emphasis on postural changes and gait (but also including gesture, facial expressions, and vocalization), as means for communication. Combined with this information-theoretic model, the taxonomy introduced in part II, and a notational scheme introduced in part III, part III answers the "*Why?*" question about movement, illustrating how humans make sense of movement in personal and individual ways—and thus how they make sense of machines.

In chapter 9, we analyze a variety of important applications with machines that have been—or could be—affected by this work. Further, we outline a process for engineers and designers, not only calling attention to how movement affects technology design, but pulling apart the process of observation, context, and prior experience and identifying how each affects how and what we see. The chapter also connects this analysis to examples of HRI research in this area.

With context and fodder from these application areas, chapter 10 delves into the complexity of notating movement, applying the symbology and principles from parts I and II in a series of examples that also highlight the relationship between a particular taxonomy and what can be observed (and thus notated), offering engaged readers more opportunities for practice. Finally, the conclusion of this book broadens our scope to consider numerous somatic and choreographic lenses from a collection of movement traditions and practices across the experiences of many bodies.

We end with a nod to the ongoing explorations in movement practice and theory that allow the kind of design process that has been used to produce pleasant physical spaces—like the Downtown Mall in Charlottesville, Virginia, where the perspectives in the prelude took place—and that will be crucial to creating harmonious environments that contain both natural and artificial movement for both humans and robots. Such facile design is enabled by systematic understanding of the medium of movement. We establish a particular approach that we are calling "the BESST System" as our primary tool, but we also offer glances toward other movement practices, including the design of movement-based algorithms, machines, and devices, as a key source of such knowledge. As humans that successfully navigate a wildly dynamic and complex planet, this knowledge is in all of us, and our book aims to write some of it down.

"Dancing, Happy Robots" (What This Book Is *Not* About)

Throughout this introduction, we have stressed the malleable, unfixed nature of movement and the meaning made through it. Consider the first version of the Nest Protect smoke detector, which was silenced by the wave of a hand. Its designers' intention was to create a system that could be easily turned off for errant cooking activities that annoyingly trigger smoke alarms, but because these designers had little training in how humans make meaning from postural and gestural shifts in movement, they failed to consider that waving arms can also, in the right context, indicate the need for help. This oversight led to the device's recall (Skybetter, 2016). A wave is an action that means different things in different contexts; moreover, a hand extending from side to side in an arcing, repetitive pattern is a movement that can be identified as many things: a wave, a wipe, a shoo, a stir. This is even true of language and stereotyped gestures like a "thumbs up" sign, which can be used rhetorically or sarcastically, drastically changing the meaning of a common symbol employed by some cultures in some time periods.

Words (and other symbols) change their meaning in new contexts and when spoken by particular speakers. Movement is the medium for all these symbols, which must be recognized, repeated, and understood on distinct bodies across distinct contexts. Thus, this book will not provide a lexicon[4] for *what movements mean*. We are passionate about this point. While it is possible to construct scientific experiments that measure similar responses to the same stimuli across a particular population, this is not the process of movement analysis, and it creates a quagmire for automation, which may be deployed as a diagnostic tool in medicine, surveillance, advertising, and finance. Already, tools in these spaces that attempt to automatically determine the meaning of human behavior (like the Nest Protect) have been identified as biased and faulty. Such effects are magnified for marginalized communities that are underrepresented in the design process, not well represented in data sets, and more vulnerable to the repercussions of misidentification.

Making a robot with motion that is meaningful to its users may have as much in common with the process of writing a best-selling novel as designing a structurally sound bridge. Dictionaries of languages, which are specific to time, place, and culture, include multiple meanings created, edited, and refined on an ongoing basis by human experts; experts of language debate the meaning of important texts for decades; and automated speech recognition is limited to low-stakes, simple applications in narrow contexts (e.g., automated call centers). This is the approach that we motivate in this book. Every movement of a machine must involve a painstaking degree of user and designer input, refinement, reconsideration, and adjustment from human experts—and must be tightly coupled to a particular user and use case.

We will briefly share research[5] that highlights how affective and valued judgments of movement, such as "happy" and "beautiful," break down across contexts. This does not mean that we cannot make artificial motion that is read by a large pool of human subjects within a particular experimental setup as "happy" or "sad"; indeed, we will also discuss research showing that such a feat is possible too. Instead, we aim to illuminate for readers that creating a successful piece of technology is much more similar to creating a work of art than they may realize—especially when that piece of technology moves. Instead of encoding meaning into motion, we impart the practice of considering *when*, *where*, and *for whom* a particular meaning may manifest. Moreover, these meanings need not be limited to topics of affect, mood, or emotion; sometimes a movement is just about scratching an itch.

This is counter to the approach of many research groups, including some of our own collaborators (e.g., Knight & Simmons, 2014; Zhou & Dragan,

2018; Luo et al., 2020), and we hope for spirited academic debate to continue on this point. Such debate has lived within the RAD Lab, creating important research insights. For example, Umer's work on gait grew out of his ongoing insistence (despite resistance from Amy) that it would be an interesting domain. And a debate about the viability of labeling a motion pattern as "happy" inspired a series of studies that reveal the benefit of explicitly modeling context in predicting human labeling of motion style (Heimerdinger & LaViers, 2019).

Likewise, the ultimate end of this book is not to design "dancing robots." In fact, we hope that the book will broaden and complicate your notion of what "dance" and "robots" are. Although you can certainly look up formalized taxonomies of movement, like ballet and tai chi, that proffer certain steps as being part of a tradition, often with associated meaning, in this book, we use the field of dance as an intellectual pursuit, a form of research, not a categorization of movement. As such, there will not be a section of the book that lists movements that are dance and movements that are not (as we are most interested in the *field* of dance, a knowledge base that helps understand all movement, instead of the social *activity* of dancing).

We will not be specifying how to manufacture machines that do what humans do; rather, we hope to teach you why you think that machines, which are still vastly dissimilar from any human form, do anything vaguely similar to what you do at all. Namely, that is, *the dance is in you*, so you observe it with other entities. Likewise, this perspective and the detailed focus on sensing and feeling our human anatomy motivated in this book—which especially emphasizes the human spine and viscera, which are often overlooked in favor of the function of distal joints in robotics and biomechanics—put into question the notion of what constitutes a "humanoid." In fact, in some contexts, a well-designed expanding and contracting faceless orb may have more in common with human motion, as it captures an abstract notion of breath, than does a back-flipping humanoid.[6]

Moreover, we want to challenge the idea that a "step" is ever performed the same way, instead emphasizing the wide variety of shapes and sizes that bodies (both natural and artificial) come in and the unique experience and variability that seem to be inherent in human movement. The point of this book is to provide readers with a descriptive (rather than an evaluative) taxonomy with which to describe your inherently personal observations of movement. Such a conversation is especially important as we look toward creating systems that function correctly across populations in an inclusive manner. As two people with particular bodies authoring this book, the way that we see movement (and, likewise, the way our movement is seen) are not universal. Becoming aware of this individuality is an awakening that can be dealt with only by including anecdotal, first-person experience (often through qualitative and subjective methods).

This relationship between one's experience of moving and one's perception of it extends to our approach to defining machines. We do not spend much time differentiating among models to describe human movement (e.g., an algorithm that detects and labels distinct styles of human gait from motion-capture data) and models to execute complex movement phrases (e.g., a bipedal device that creates various styles of gait recognizable by humans). The former example concerns sensing and interpretation, and the latter example concerns actuation and generation. But these are two sides of the same coin: both will depend on how we define, construct, and deconstruct gait. We have to contend with that interdependence, especially in gross body movement like posture, gesture, or gait. Typically, we will discuss movement in the context of human bodies because that is the source of how we see patterns in movement, but we will do so with a taxonomy that is amenable to abstraction, and thus translation to artificial bodies.

An Introduction to the Bodies That Wrote This Book

On a rainy day in the fall of 2013, the two of us met over pastries at a coffee shop under the vague pretext of "this is someone you should know"—connected through mutual acquaintances who noticed two professionals for whom the daily movement practices of their youth had grown into professional obsessions with embodied research. Many young children engage in such practices daily or weekly through sports, dance, and recreation, but as two adults who thrived on dissecting movement into parts and synthesizing those parts as a primary mode of meaning-making, we stood out as kindred spirits in the small community of Charlottesville, Virginia. Upon meeting, we found other shared aspects of our experience: our training in ballet, contemporary dance, and yoga. Moreover, we share privileges: we are both petite, white, cisgendered, heterosexual, able-bodied married women with sufficient economic security to allow us to engage in work that often does not pay.

We came into a deeper relationship when one of us (Amy) began studying with the other (Cat) at the Laban/Bartenieff Institute of Movement Studies (LIMS) to become a certified movement analyst. Cat was an affiliate faculty member and coordinator of movement analysis certification programs with the institute and has used the material in her work in dance (both in the classroom and in her artistic practice) for nearly four decades. Amy was just starting her own robotics lab at the University of Virginia—one that, from its outset, was designed to combine her experience in dance and engineering to create more expressive robotic systems. Movement studies was something that Amy had been leveraging during her doctoral studies, like a growing number of roboticists, searching for a way to study movement such that she could program machines to move in ways that could be salient for humans. The motivation for this emerging intersection of work is that such profiles will be a critical component of interactions between robots and humans as robots move into human-facing scenarios. Thus, completing the two-year modular training program, which Amy would soon begin in Cat's studio, was at the top of her agenda as a new professor.

The studio settings in which much of the field of movement studies has developed inform challenges in disseminating knowledge to academic communities and are inherited from the system's founders: Rudolf Laban (1879–1958), a choreographer and artist, and Irmgard Bartenieff (1900–1981), a dancer and physical therapist. Likewise, the practice of movement techniques heavily used in the performing arts, like those of F. Matthias Alexander (1869–1955), Martha Graham (1894–1991), Merce Cunningham (1919–2009), and William Forsythe (b. 1949), find the primary site of their transmission inside the studio—and theater—rather than in a book.

Laban's interest in movement came from a life in the arts, mysticism, and choreographic practice. He studied the movement of his dancers and worked to create a system for organizing and describing it. His most famous contribution is the invention of Labanotation, a system for movement notation that is still used today, including for the practice of copyrighting choreography. He is also well known for his creation of harmonic movement scales (analogous to musical scales in different keys), and his studies of factory workers during the Industrial Revolution (where he used his dance-inspired principles to improve the functional efficiency of their workday).

Bartenieff's introduction to the system came first as a dancer and a student of Laban. She then developed a successful practice as a physical therapist, working with both polio patients and professional dancers. One of her key innovations was in developing exercises that helped both seemingly disparate groups improve their access to effective movement patterns. These bodily exercises came from a more sensorial, interior place than Laban's work in stagecraft and performance, and her most prominent contribution to the system was in growing the somatic, body-based aspect of the work.

The work of Laban and Bartenieff in developing theory occurred in parallel to a shift in the practice of Western dance. Traditional classical ballet tightly coupled itself to a rigid bodily vocabulary but was challenged in the twentieth century by expressive pioneers like Cunningham, Graham, Loie Fuller, Isadora Duncan, Katherine Dunham, Doris Humphrey, José Limón, Alvin Ailey, Yvonne Rainer, and, more recently, by Forsythe, Bill T. Jones, Twyla Tharp, Mark Morris, and Ohad Naharin. Their work formed modern, postmodern, and then contemporary dance, each of which came with their own forms of physical training and modes of expression. We, Amy and Cat, are products of that evolution, centered in these particular, Western ways of thinking about movement and using it to express oneself.

Like many other dance artists and teachers, Cat was drawn to the Laban and Bartenieff work in the 1980s as a way to formalize her practice in teaching dance to university students (she founded the dance program at Drew University around the same time) and in creating work with her professional dance company in New York (Offspring Dance Company), while also completing the Certification in Movement Analysis (CMA) program at LIMS in 1984. Amy was attracted to Laban's and Bartenieff's ideas as part of her research in the late 2000s (at Princeton University) and early 2010s (at Georgia Tech and Emory University). As many dance students do, Amy encountered them throughout her dance training, often informally, in classes designed around Laban's taxonomy of body actions or in warm-up routines incorporating Bartenieff's famous exercises. As a researcher, she studied their work from books and papers aimed at quantifying elements of it for application. After completing the CMA program at LIMS in 2016, her understanding of this work became rooted in her own unique physical experience, complicating what was an initially straightforward idea: translating movement to machines.

Parallel to her studies in engineering, Amy bolstered years of dance training through exposure to physical, bodily rigor that had a new emphasis: description through systematic, symbolic notation. Although she had spent years in the studio—like many soon-to-be movement analysts—beginning as a young child, it was not until college that she was asked to critically reflect on dance as a practice and medium for communication. In graduate school, she participated in the Fieldwork program, hosted by the New York-based arts organization The Field, where artists use a particular, structured style of peer-to-peer feedback that eschews valued judgments for reflective descriptions to better help artists find their own desired expressions. Her training at LIMS further deepened these approaches to movement, providing her with new terminology to describe human movement and a more rigorous framework for examining the process of making meaning from it. The training also helped express intuitive ideas about movement into something that could be written and reasoned about with others in the process of developing new technology.

The Value of Personal, Bodily Experience

Although this book is one of the first to address the application of movement studies to machines, we are one of many to include an experiential perspective of movement in the design process. Designers of products from vegetable peelers to furniture, and environments from the interior of private homes to public parks, have also employed this approach. In fact, the outdoor promenade in Charlottesville, Virginia, where the personal perspectives in the prelude were written and our collaboration began, is a perfect example of a space shaped by the influence of a dancer's expertise.

The downtown mall was designed by the acclaimed landscape architect Lawrence Halprin, who also created Ghirardelli Square in San Francisco, in concert with a technique for observing movement developed with his wife, the renowned dance artist Anna Halprin (Worth & Poynor, 2018). This process, which included the creation of movement scores (Halprin, 1975) and participation from many community stakeholders (Halprin, 2014), resulted in many distinctive features of this mall (Hirsch, 2011). For example, rather than dotting the space in linear, predictable lines, irregularly placed trees cause walkers to dodge and weave along the mall, encouraging more meandering pathways of travel for occupants of the space. In the context of this promenade, such patterns of motion may be associated with relaxation and recuperation, whereas linear, direct pathways might have been associated with expediency and exertion. It creates a pleasant and distinctive space that offers views of the Blue Ridge Mountains and facilitates successful businesses, public events, and community-building in the downtown area.

Such a process is not so different from the one that we encourage in this book, paying explicit attention to experience, movement, and human perception of them. If Lawrence Halprin merely noted the kind of experience he wanted to create and not distinct pathways for walking in the space, or if he had merely written the options of the distinct pathways for occupants and not what these two choices might mean in the context of a pedestrian promenade, he would not have ended up connecting mechanisms of movement to the experience of particular architectural features, likely failing to make the same successful design. Collaboration with dance artist Anna Halprin and the use of movement scores (notation) facilitated this needed connection.

Technology designers are increasingly thinking about physical interaction between humans and artificial agents in shared spaces. At the 2019 US National Robotics Roadmapping Workshop in Chicago, Illinois, employees from Amazon and FedEx described their challenges with warehouse robots. An anecdote from one employee, who rearranged her path of travel to avoid the autonomous robots because she feared them, saying that she could never tell what they would do next, paints a vivid picture. Even while the safety systems on these devices ensured that the machines did not pose a real physical threat, workers faced confusing path-planning choices of their own when faced with opaque, monolithic devices for which designers did not create conventions in observable changes to help employees understand the internal state of those devices. These particular robotic challenges are choreographic and require nuanced observation of humans' somatic experience to resolve.

A key difficulty in design problems like the ones formulated by the Halprins, Amazon, and FedEx is the ubiquity of movement. Just like air, or the blue of the sky, or unconscious bias, movement, through its ever-presence, becomes hard to notice—and often even harder to define. What constitutes a "movement"? How is it measured?

In this book, we will posit that the answers to these questions are not yet known, but what is clear is that the answers comprise a perceptual phenomenon. All matter is constantly in motion; it is the salient moments where a "movement," or several "movements," are *consciously perceived* that interest us in this book. Somatic strategies—the application and development of somatic practices for producing and observing one's own bodily movement—help illuminate our personal experience of movement. Choreographic technologies—tools that emerge from the practice of making and sharing dances—provide design methods that align with the perception of human audiences. Notational abstractions—specific systems for representing movement—connect these perspectives in forcing concrete observation and creating opportunities for reinterpretation and translation. And so, just as we were able to sing before we could notate music and tell stories before we could write books, the best way to formalize the knowledge that comes from embodiment is to look within, move, and dance.

I Making Meaning through Movement

1 Noticing Movement: Meaning, Measurement, and Experience

This chapter provides a wide overview of meaning-making and movement. **Movement studies,** including **movement analysis,** discussed in chapter 2, provide an intellectual landscape with which to understand how humans construct (and likewise deconstruct) movement. In chapter 1, we take a broader focus, reviewing many disparate fields to better situate chapter 2: philosophy helps us understand how movement translates into meaning; engineering provides quantitative models and empirical measurements of bodily movement; and phenomenology gives us a framework for understanding how the felt experience of the body relates to movement perception. Crucially, these fields create a picture of how humans perceive themselves in their own observations about the world. This is not so different from machines: the same movement model used to generate robotic behavior through actuation can also be used to interpret the behavior of counterparts through sensing. For humans, since our perception depends on our own unique bodily experience, the meaning that we make from our sensations is unique, personal, cultural, contextual, and tied to the human form, casting anthropomorphism—and any sense of shared meaning with a foreign body—as a property of the observer as much as (if not more than) the observed.

1.1 Meaning-making
The personal process of constructing salience from elements in our environment

Phenomenology brings an intellectual framework to conscious experience. Maurice Merleau-Ponty studied experience and perception in the process of making meaning. He centered the subjective, qualitative nature inherent to human attention and information processing after noting how his ability to perceive (or overlook) detailed features about an event or object changed the meanings of these observations. For example, Merleau-Ponty (1945, p. 5) contended that the meaning of a red carpet changes if it is noted that it is a thick, wooly carpet as opposed to a smooth carpet with low pile. His work grapples with how to disentangle ourselves from the world in which we are constantly interpreting our sensations—making meaning—in order to understand the very thing that we are trying to perceive. In other words, our subjective experience is embedded in everything we think we understand; we are constantly imperfectly perceiving, and in doing so, changing both our access to the world and our experience of it.

We see this in the body very literally: an exercise in finding a centered, neutral posture will often engage participants in a bouncing and shifting of their mass to feel different distributions of weight (say, in the feet). Each bounce and shift changes what the mover is sensing and what the body's internal state is; thus, when returning to a previous posture, that posture feels different. Coming to a resolution about where one is centered and neutral in posture is then an ever-shifting compromise and estimation, changing moment to moment and day to day. On a day when the feet are cold and stiff, "center" may feel like a narrow regime of tenuous balance, whereas on a day when the feet are warm and pliable, "center" may feel like a wide space of play and activity because the bandwidth of perception and action in the feet is greater when they are enlivened and "awake." You can experience this by grabbing one foot and giving it a thorough massage, heating your foot with the action of your hands; then try balancing on that foot and the other, unmassaged foot in turn.

The interconnection with training and perception is also observed in scientific studies of human perception. For example, olfaction and the human ability to evaluate and describe olfactory sensations are found to differ distinctly across cultures. In a study comparing Jahai hunter-gatherers from the Malay Peninsula and Dutch participants from the Netherlands, Majid et al. (2018) discovered that the Jahai are more adept in expressing their experience of odors in language. The researchers noted faster response times and the use of more dedicated, abstract language in Jahai participants compared to their Dutch counterparts, confirming previous studies that showed that other hunter-gatherer cultures are better at describing odors (Majid & Kruspe, 2018). Just as wine sommeliers distinguish among subtle variations in wines, even noting the differences in taste between vintages, this research suggests more broadly that prior training affects what we can externalize about our experience and that language is a nonunique abstraction that can be better or worse at describing our sensations. We experience similarities in working with movement: the dedicated taxonomy presented in part II equips movement analysts to more easily, accurately, and rapidly describe movement.

The philosopher Maxine Sheets-Johnstone (2011, p. 438) writes that "the primacy of movement" makes it hard to examine directly, and language is "post-kinetic," or a behavior that is a result of our movement. Language is movement, and many other things that are not language are also movement. Thus, it can be confounding to compare meaning-making with written language to meaning-making with movement. The notion of a movement dictionary relies on limited contexts and formalizations of movement. For example, Raindel et al. (2021) study movement associated with "dramatic action and emotion" in order to develop such a lexicon to link body postures and their meanings, finding that descriptive verbs were stronger labels than emotional states.

Emotion is one way that we assign meaning to movement, and trained actors use many cultural conventions in communicating such states through their movement (Goodman, 1976). But it is easy to imagine moments where the same posture takes on entirely different meanings: a clenched fist thrust into the air can indicate anger, victory, or having just successfully squashed a mosquito. Likewise, the notion of what a word "means" (in the sense of the elements contained in that word's entry in a dictionary) is distinct from what an actual, situated utterance of a word *means*. As such, this book will not codify stereotyped expressive actions—a thumbs-up, for example—and list their meanings; it is instead concerned with the embodied process of coming to meaning with foreign bodies.

In studying language, linguists deal with meaning through the notion of symbols, and the process of symbolic representation is often linked to language and language formation by psychologists who study language acquisition and development. Scholars Alex Gillespie and Tania Zittoun (2013) heighten the role of the body and its movement (instead of emphasizing language) in meaning-making. They use the example of a family with an infant and a toddler to describe how symbols form, explaining that when the baby cries, the older sibling does not know the meaning of this bodily action or what will come from employing it. But the older sibling observes the reaction of the parents to the baby's cries and learns to associate meaning with it. The older sibling sees the parents jumping to give attention and care to the newborn—and is even herself drawn into the act, creating a relationship between the baby and the rest of the family. In fact, the next time the older sibling cries, she does so with this new knowledge that crying will elicit a caregiving response from her parents. This example highlights how meaning is created and reinforced through lived experience that is specific to one individual.

Consider a similar example with technology: computer developers trying out a new device may poke at a button that they do not know the purpose of, developing a sense of meaning through their observations of the device's reaction, whereas lay users might be socially conditioned away from pressing unfamiliar buttons. Gillespie and Zittoun (2013, p. 528) assert this idea of meaning's mutability and contextual nature explicitly:

> Symbols arise at the points where internal, personal, embodied and emotional experiences meet an external social or semiotic structure. Meaning is where personal sense and shared meaning meet. The personal sense comes from our own unique embodied trajectories. Although our experiences are socially determined, that determination works on our own individual bodies, stirring individual emotions, creating personal sense. This personal sense finds expression, or resonance, in social settings and semiotic structures. Equally, these institutional settings and semiotic structures need to find the relevant personal sense, the relevant past experiences and embodied memories in their participants, to function.

This work extends prior semiotic theory on how meaning is made through symbols and how symbols are acquired through *context* (Werner and Kaplan, 1984). It is possible to see this idea in terms of how art becomes meaningful as well, noting especially that the meaning of any particular piece of art requires contextual cues and cultural context; changes over time; differs for each viewer; and may wildly differ from the intention of the artist. For example, one could fall in love with a painting and buy it in their twenties and ten years later—after accruing new lived experience and shifts in societal and cultural norms—see the painting in a new light, maybe even feeling the need to get rid of it.

The phenomenon of shifting meaning motivated postmodern philosophers and artists as well. A philosophy developed by Jacques Derrida (1974), *deconstruction*, purports that a text (or work of art) cannot have a fixed meaning, and indeed whatever meaning is made also changes over time. This speaks to our understanding of meaning-making as contextual, based on prior experience, personal, and unique, yet able to be *constructed* by recognizing this mutability and making one's own choices. We invite the reader to consider movement, the body, and the practice of embodiment as a process similar to construction and deconstruction that will support and inform design and making meaning with machines in motion (through exercises introduced in chapter 3 and used throughout part II). You have to move your body to understand what you perceive, but this movement (this observation) also changes what you perceive, as seen in the exercise for finding centered posture described at the beginning of this section.

Because meaning is individual and unique to each of us, each observer of the same moment can arrive at a wholly different meaning. This is most simply because each of us is a unique body brought to that moment by a unique series of movements and making sense of it through the movements available to us. These prior movements form every individual's experience, comprising training, culture, personality, and other characteristics, as well as a unique physical form created through innate biological composition and a series of interactions with the environment during development. The meaning that any individual perceives in the moment is a direct function of that prior corpus of exposure and bodily experience.

1.2   Movement as Quantitative, Objective Events
Measuring movement events, often with technology

If meaning comes from movement, what is movement? Can we measure it? And, if so, what are the units of measure? Can two distinct bodies, perhaps with different physical dimensions, perform the same movement? How do we record a movement event? And what do such records leave out? What does it mean to perform a movement again? What is *not* a movement? In the context of considering meaning-making as a process achieved through motion that was set up in section 1.1, questions like these complicate the notion of measuring and recording movement events and reveal the subjectivity inherent to such an act.

1.2.1   Quantitative Movement Records
Recording movement has been a subject of interest since the first cave paintings, in which early humans scrawled static snapshots of themselves

and animals performing activities in their environment. And while movement must be inferred in these renderings, the drawings do convey a sense
of the dynamic actions and interactions with the environment. During
the Renaissance, artists developed a new understanding of the relationship
between a two-dimensional canvas and the three-dimensional world. Thus,
although they remain inherently subjective renderings that reflect a particular artist's experience, drawings use different innovations for capturing
the detail of such an interaction with both concrete and abstract visual
representations of the world (Dickerman & Affron, 2012).

Early experiments at quantifying human movement as objective events
often employed photography, as re-created in figure 1.1, using long exposures to illuminate the pathways of various body parts, particularly during
work. In the early twentieth century, Nikolai Bernstein (1926/1967) studied
movement during manual labor, postulating that acts like the one shown
at left in figure 1.1 are composed of many smaller movements. He thought
that understanding these smaller movements could optimize movement's
efficiency during manual labor. Frank and Lillian Gilbreth's (1919) applied
motion studies used similar techniques around the same time as Bernstein's
original work, although their approach—attaching lighted elements to
human appendages—seems to emphasize more variability and chaos in the
revealed motion than Bernstein's neat cyclograms. Eventually, Eadweard
Muybridge's (1887) studies taking rapid-fire sets of sequential images of
horse gaits would initiate the development of motion pictures and reveal

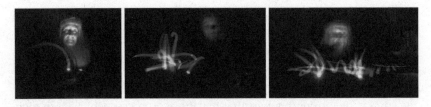

Figure 1.1
Author recreation of early measurements of human behavior offering an external-
ized view of movement. Left (Cat): Similar to Bernstein's (1926) cyclograms, this two-
second exposure shows the motion of one arm in a single action, revealing the smooth
coordination of human motion. Center and right: Similar to Gilbreth and Gilbreth's
(1919) motion studies, these eight-second exposures show the motion of a longer
phrase, revealing the chaotic variability of human motion. The image in the center
(Amy) shows one arm, and the image at the right (Cat) shows the action of two arms.

much about natural motion (specifically, he was interested in whether all four hooves of the horse left the ground at the same time), as discussed further in section 10.1 of chapter 10.

Video opened up new realms for movement analysis, as well as archival records of movement, and probably facilitated a shift from experiencing movement as only a fleeting dynamic process that is felt inside the body to one that can be captured in another medium, documented, and shared. Eventually, the quest for richer external data sets of movement and advances in pattern detection algorithms led to three-dimensional capture techniques. Today, researchers have access to both optical motion-capture systems that detect reflective markers or other vision-based landmarks, as well as accelerometer-based active systems that detect the physical movement of limbs. Software can then approximate rigid skeletal models to explain those sensor readings as postures with associated time stamps. Typically, for full-body motion studies, researchers use skeletons with around 50 degrees of freedom—points of movement (or "joints") between rigid bodies (or "bones")—to model human motion, but filmmakers often use more detailed models to capture the nuanced facial and muscular expressions of movement actors and translate this action into animated characters through a labor-intensive process.

These developments reflect a massive progression of machinery and technology to detect human motion, which we do not aim to detail here. However, it is instructive to consider an example of the kind of motion record that motion-capture technologies extract. Figure 1.2 shows an example of a three-dimensional, room-sized capture space, and figure 10.2 in chapter 10 outlines one of the original data formats that emerged for documenting motion capture data. This is given to provide the reader with a mental image of motion capture, which is often used as a form of ground truth for human motion and is even leveraged in dance performances sometimes (Dils, 2002). This data captures quite a lot of individualism and recognizability. It harkens to motion studies with points of light that occlude most of the human form, which is nonetheless surprisingly recognizable to subjects (Vanrie & Verfaillie, 2004), including when using high-level, emotive labels (Clarke et al., 2005). We see a similar phenomenon in motion-capture data: it can include the distal edges of a body's motion while typically not detecting the motion of the fleshy torso, face, and musculature; nevertheless, the systems record recognizable motion.

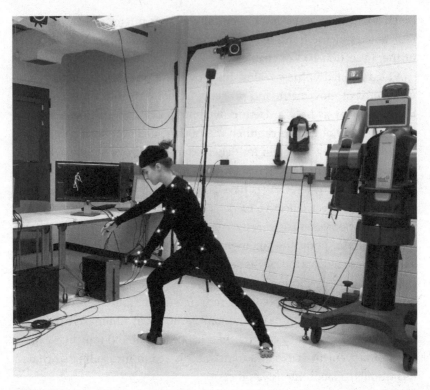

Figure 1.2
Motion capture with passive, reflective markers in the RAD Lab at the University
of Illinois Urbana-Champaign. The computer screen displays the three-dimensional
model of movement extracted by specialized software from data collected by the
overhead cameras that record infrared reflections from the wearable markers on the
subject (Amy). Even this motion model lacks many essential details of human movement. An example of a resulting data structure is shown in figure 10.2 in chapter 10.

We are especially interested in noting the shift that machines have
caused in how we study and conceive of the process of moving. Being able
to see one's own movement *outside* the containers of our bodies is a marked
intellectual and experiential shift. In some ways, it afforded researchers in
labs like that of the Gilbreths the kind of perspective that a rehearsal director might have with a dance company: watching bodies from an external
standpoint for analysis. But it also goes deeper than that—first, by offering
options such as rewind, which allow movement phenomena to be viewed
in great detail; and second, by suggesting that movement is a measurable,

objective event. In another sense, however, these recording techniques may be seen simply as an extension of cave drawings, still relying on the subjective perspective of the operator.

1.2.2 Trajectory-Based Modeling

Early measurements drove the development of mathematical models designed to capture and predict how humans moved. In most cases of motion capture, energetic measures of force such as acceleration, velocity, and displacement are employed, which focus specifically on distal bodily limbs. This is evident even in the first studies shown in figure 1.1, where Bernstein was working to optimize human labor under funding from the Soviet Union, establishing a trend that continued in subsequent research, such as Flash and Hogan's (1985) seminal study. Measuring the trajectory of human arm movements, they postulated a quantitative model purporting that humans were selecting minimum jerk pathways, or pathways that minimized changes in acceleration. Like Bernstein, Flash and Hogan used optimization to explain their observations and measurements of the distal joints of the human form in motion.

Trajectory-based metrics are often the basis for comparisons between robots and humans, as well as forming the architecture for *producing* robot motion. Thus, it should perhaps come as little surprise that robots have outpaced human counterparts in the range of magnitudes of acceleration, velocity, and displacement that can be achieved on these platforms. Moreover, robots excel at repetition and repeatability of action, something that Bernstein was also very interested in: he wanted workers who could strike a hammer evenly and repeatedly. A similar idea of ideal workers was used in Karel Čapek's play *Rossum's Universal Robots* (1923). The Czech playwright imagined workers that did not get distracted or need to take breaks, coining the term "robot."

What robots *still* do not do well, however, is create complex patterns that solve tasks or express meanings in dynamic environments. This kind of behavior is common in humans and has been studied extensively. For example, Del Vecchio et al. (2003) examined humans drawing simple figures in a computer program (Microsoft Paint) where users drew lines using a handheld mouse. In their setup, the researchers monitored the resultant actions of the subjects—how the cursor evolved on the screen—and worked to classify the types of drawing tasks that emerged over time. Using second-order ordinary differential equations, which captured the patterns of acceleration and velocity in the resulting lines deposited on screen by their human subjects, the researchers identified distinctions like "pickups," when subjects stopped drawing to move to another area on the screen, as well as "straight" and "curved" lines, distinguishing when subjects drew the two line styles. Related and follow-on work has explored this idea in similar ways as Del Vecchio et al. (2003): through the notion of *movemes* (Bregler, 1997) or *movement primitives* (Fanti, 2008). Further discussion of using motion primitives, particularly in the context of robotics and human-robot interactions, appears in chapter 9.

Studying movement in humans by fitting empirical models to external measurement is one way to try to understand the categories of our actions, and subsequently build models for robotic systems. However, we could also imagine abstractions that are not quantitative and objective, but instead are qualitative and subjective. Such a categorization—one that centers on the physical experience of the body—is how pitches came to take a central role in the notation of music, which we will discuss further in chapter 2. However, the quantitative structure of these ideas (i.e., frequencies corresponding to each note) was figured out only many years later.

1.2.3 Informatic-Based Modeling

Another way to look at movement is through the lens of *entropic* measures, a way of modeling that attempts to consider how much information movement contains. Ofli et al. (2014) attempted to parameterize joint angles that are "most informative" using statistical notions; the team used the fact that joint angles with more variance in a given activity are more informative than others, which led to improvements in human activity recognition. Berrueta et al. (2018) similarly approached the topic through the lens of functional tasks. In line with a notion of dynamical movement primitives used in Del Vecchio et al. (2003), this team considered the relative entropy between two constructions of a task, leading to an assistive device that better aids humans in physical collaboration.

At the same time, research influenced by our collaboration has suggested that all movements, in the right context, can be equally informative (LaViers, 2019a). This approach creates instead an information-theoretic measure of distinct **movement platforms.** In this model, different platforms are more or less expressive based on their morphologies (because larger or smaller numbers of postural shapes are available to each combination of hardware and software). By counting the number of possible postures based on the physical characteristics of the robot design, looking at the range and resolution of its actuation and sensing systems, this work assigns a number of *bits*, modeling the capacity of each platform as an information source (a concept discussed further in chapters 7 and 9).

In addition to objective measures of information content, researchers have examined subjective measures. Famously, Heider and Simmel's (1944) work with simple abstract shapes (such as triangles, squares, and circles) moving around on a blank surface showed that human subjects can read consistent and similar narratives from the movements of foreign bodies. Reeves and Nass (1996) proposed the "media equation" to describe how people across a wide variety of experiments that they conducted in the 1990s treated artificial representations as real-world entities (e.g., thinking that popcorn shown on a television screen would fall out if the set were rotated upside down and exhibiting politeness toward computers). Similar results have been produced more recently through direct surveys of human observants; for instance, in Darling et al. (2015), the subjects were less likely to be willing to smash robots for which they demonstrated empathy, which was especially heightened through shared narratives. Gielniak et al. (2011) showed the benefit in human perception of robot motion when random noise (highly entropic movement) is added to the motion, causing human viewers to imitate the action more easily. We say that this kind of research looks through the lens of *expression* rather than *function*,[1] as it explicitly considers human perception of moving bodies.

1.2.4 Language-Based Modeling

Language is a key part of culture and communication, and movement is the medium that allows each body to form symbols from this shared corpus through vocalization (and written word). Thus, it is not intellectually rigorous to say that movement is a language; rather, movement subsumes language (Sheets-Johnstone, 2011). In fact, all language is created through bodily movement, both languages that utilize vocalization (e.g., spoken English) and those that do not (e.g., American Sign Language). This is one reason why the concept of "body language" is misleading. Researchers dividing a human subject's posture and gesture from facial expression showed that "body cues" are better than facial expressions at predicting how a person's movement will be understood, in terms of positive or negative valence of emotion, by a human observer (Aviezer et al., 2012). The system for movement analysis presented in this book will also focus on gross bodily movement, but the theory driving it, like that of Sheets-Johnstone (2011), presumes that body movement is a broad activity that includes vocalization, facial expression, postural shifts, patterns in breathing, and other elements. Moreover, we will emphasize the crucial role of context and the environment in understanding the varieties of possible meanings in bodily motion, especially when looking beyond the relatively narrow set of movement behavior that produces verbal or written language or facial expression.

Nevertheless, prior work bridging movement studies and quantitative modeling has used transition systems like those that describe the structure of languages (both natural and computer programming languages), as shown in figure 1.3, to model artificial and natural motion. Modeling the behavior of actors portraying a sentry or guard, Gillies (2009) chopped up bits of motion-capture data and described its structure in a finite state machine, which could be used to create novel sequences in the same role. LaViers and Egerstedt (2011) used a similar structure to explain how ballet barre exercises, which are relatively short actions that are thought of as

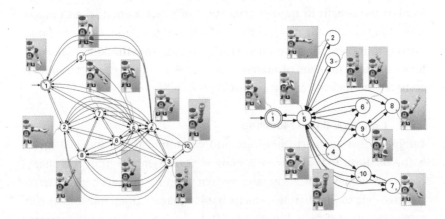

Figure 1.3
Using finite state machines to represent movement. Each state (numbered circles) corresponds to a body pose; each transition (arrows) indicates the possibility of moving between such poses. The structure on the left describes a "disco" style of motion, while the structure on the right describes a "cheerleading" style of motion. However, the trajectory between postures is not specified. From LaViers (2013).

single movements, describe longer, more fluent phrases of ballet technique, proposing a model of motion that captured only sequences of key poses, not fluid trajectories of continuous action.

Such formulations do a good job of discretizing base movement behavior and generating variations—or styles—of it through recombination. More recent work has explored trajectory-based formulations for capturing continuous aspects of style and generating varieties of motion through optimization frameworks (see LaViers & Egerstedt, 2012; Zhou & Dragan, 2018). Variations in movement execution, in both spoken language and bodily movement, create the capacity for differentiation and communication (LaViers, 2019a).

1.2.5 Subjective Validation of Events

Del Vecchio et al. (2003)'s work with drawing tasks, described in section 1.2.2, highlights an important idea: classification and segmentation go hand in hand. With a different set of template actions, the researchers would have chopped up the subjects' motion at different points in time, creating a different set of movements. Something *feels* "right" about their resultant classification, but it can be difficult to objectively justify why. For example, a trivial classification and segmentation of the activity would be "drawing" and "not drawing," lumping all of the observed activities into one type of task. Or perhaps we divide the "circle" line style into smaller parts or quadrants: north-to-east arcs, east-to-south arcs, south-to-west arcs, and west-to-north arcs. This labeling also explains the data accurately. The possibilities are truly endless; thus, subjectivity is inherent in the evaluation of their work.

Consequently, we need to address in the remainder of this chapter the issue of how subjectivity comes into play when explaining movement and meaning. For example, the model of minimum jerk proposed by Flash and Hogan (1985) has broken down across a wider variety of activities and human subjects. The classifications of Del Vechicco et al. (2003) were acceptably presented, but there is little quantitative justification for why they would be (and it is easy to propose additional models of activity that perform just as well objectively). Moreover, the information content of movement also varies based on context (in some contexts, a given body action is salient; in others, it is not). In other words, human perception plays a role in forming and validating these models, measurements, and records that is unaccounted for in our discussion so far.

There is much about movement, such as the feeling of viscera bouncing along as we jog, that is not captured by external measurement, so these measurements may have derailed our ability to understand how we experience movement, constructing and deconstructing meaning through our actions and perceptions. Moreover, the notion that video-based records capture movement is imbued with an ableist bias that privileges the *viewing* of motion. In this book, we want to explore movement from a broader perspective, leveraging all of our possible sensing modalities. Thus, we will reexamine the question "What is movement?" and, in doing so, shift from external measurement to internal experience.

1.3 Movement as a Qualitative, Subjective Experience

Describing movement experiences, often with personal reflections

The attempts to define and measure movement quantitatively and objectively presented so far have ended with a wide range of models with a large degree of subjectivity used in validating them. When is a functional task done *well*? When is an expressive action *salient*? The answers to these questions depend on the perspective of the observer and the context of the measurement. With this in mind, let us revisit the question of "what is movement?"

- move·ment: /ˈmo͞ovmənt/ noun
  - an act of *changing* physical location or position or of having this *changed*.
- mo·tion: /ˈmōSH(ə)n/ noun
  - the action or process of **moving** or being **moved**.
- mo·tion: /ˈmōSH(ə)n/ verb
  - direct or command (someone) with a **movement** of the hand or head.
- move: /mo͞ov/ verb
  - go in a specified direction or manner; *change* position.
  - *influence* or prompt (someone) to do something.
- mov·ing: /ˈmo͞oviNG/ adjective
  - in **motion**.
  - producing strong *emotion*, especially sadness or sympathy.

These definitions[2] have us chasing our tail, so to speak, as many are defined in terms of each other (see the bolded elements of the definitions in the list). What is left (see the italicized key words in the list) amounts to descriptions of change, emotion, and influence. This mirrors an idea that is often quoted in the movement studies community but can feel like an anemic truism: "Movement is change." In science and engineering, many variables change—some that seem to produce movement and some that do not. For example, changing concentrations in a chemical reaction would not prompt descriptions of motion, and certainly not of salient gestures, which is often how the term "movement" is used when describing the human body.

Many things that are not the human body also undergo movement, such as automobiles, steel I-beams, and robotic manipulators. In these cases, where relatively rigid bodies are undergoing transformations in three-dimensional space, we can more easily specify concrete ideas of movement (and stillness): change in posture, change in velocity, change in acceleration, and so on. Moreover, given our understanding of planetary science, quantum physics, and human physiology, we—and everything that we can experience—are always moving. Planets and galaxies rotate and revolve; molecules vibrate to form matter; and living organisms constantly pump blood, move chemicals, digest, respire, and use tiny movements called "saccades" to observe the world around them.

Certainly, in human bodies, "stillness" is simply an abstraction—or a description of our experience. In this book, we are not concerned with the movement of planets or molecules or diffusion in human cells. Instead, we are curious about moments of movement that are perceived by humans, occurring at particular temporal and spatial scales (not too quick or small or slow or big to perceive), that would be described by an observer as a "movement" or a "series of movements." In this context, we contend that "movement," like "stillness," is not an empirically measurable quantity, but rather a description of our experience. In other words, **movement** is *perceived* change, which frames the process of understanding movement as a choreographic challenge.

Many notable dance scholars have grappled with how choreography, dance, and performance come to have meaning for both performer and viewer, drawing on various philosophical positions, as well as lived practice. In *Reading Dancing: Bodies and Subjects in Contemporary American Dance*, Susan

Foster (1986) examines the various ways that choreography (and movement) come to have meaning through differing contexts, intentions, representations, and cultural norms. Many of her subsequent works continue this critical examination of understanding corporeality (see Foster, 2002, 2010). Sally Banes (2011), a historian and theorist of contemporary dance, writes extensively about postmodern dance in 1960s and 1970s America, highlighting active collaboration across disciplines and the shift in performance attitudes that changed how the choreography of this era resonated with audiences.

Other scholars writing about philosophical concerns, processes of making dances, cultural contexts, and choreographic relationships with technology also emphasize the personal and contextual nature of understanding movement (e.g., see Butterworth & Wildschut, 2009). Working with digital technology in performance, Susan Kozel (2007) emphasizes the role of technology in human experience as a way to understand bodily movement, using methodologies developed through a vigorous practice using machines in performance to delve into human experience with extant technologies, reframing this relationship through the lens of performance and phenomenology. In doing so, she casts a wide net around who and what might be receptive bodies for affect and performance, addressing future-looking notions of self through our relationship with machines.

These writings also represent a philosophical shift that runs counter to the idea of measurement. Nicolás Salazar Sutil (2015, p. 237) works to resolve the quantitative, objective models of movement used in machines with the qualitative, subjective expression of movement in humans in his treatise on movement representation, writing that "we move in order to know our inner selves from the outside." Kozel (2007, p. 269) expands the definition of a body in describing her lived experience, asserting that "my data is not radically discontinuous from myself." While digital records of movement measurement (or data) have facilitated enormous insights about the body as viewed through an external lens, scholars like Sheets-Johnstone, Kozel, and Sutil recenter the experience *from within the body* in their attempts to understand it.

Despite the focus on connecting inside to outside that is described here, this book does not explicitly deal with affect, and the taxonomy presented does not use affective terms. Rather than presenting an extension of affect theory (e.g., Russell, 1980), we encourage the development of the associated images and feelings personal to each mover. Movement can be about emotion (Shafir et al., 2016), but our inner (and outer) experiences are not always laden with affect—and, even more, we often work to hide our feelings through choreographic choices in movement.

This shift prepares us to accept some anecdotes presented in chapter 2 that highlight how much of the material in this book was not learned by reading a book—but instead was developed by moving a body. This bolsters a notion of **artificial embodiment** in machines, a description of physical capabilities for machines that are complementary to, and inspired by, human (natural) embodiment. This term uses the modifier "artificial" in the same way that it is used in the term "artificial intelligence," and it encourages the idea that the materiality of external movement in the human body has dimensions, textures, and capacities that are distinct from machines. A better appreciation for the expansiveness of the human experience of movement may better guide investment in machine development as well, avoiding underestimates of technology readiness. For example, the development of self-driving cars has not played out as expected, with many companies failing to deliver on full autonomy as promised (Mims, 2021). Thus, there are a number of objective impacts of experiential analysis.

1.4   Kinesthetic Attunement through the Act of Recording
Discerning new aspects of movement through informal notation

Most of the scholars mentioned in this chapter are interested in understanding human experience, which they approach through a wide variety of intellectual frameworks ranging from philosophy to experimental methods. But what does Kozel mean when she describes intercorporeality between bodies? And what, exactly, is the bodily movement that Sheetz-Johnstone refers to, which exists before language, before we write down ideas? This is a physical, lived practice in the body, which we must find a way to access through the pages of this text. A concept that may help is something called **kinesthetic attunement.** This term is used in multiple communities of studio-based dance practice, but it shows up occasionally in published academic writings, particularly in the context of dance (Gray, 2015), therapy (Samaritter & Payne, 2017), and especially Dance/Movement Therapy (DMT) (Young, 2017).

We can choose to come into a deeper noticing of the movement of ourselves and others through kinesthetic attunement. In this process, we take the movement of an observed body into our own. It can be as simple as mimicking someone else's gait (an exercise that we invite the reader to try at the end of chapter 3) or as elaborate as rewatching the same piece of choreography over and over in order to learn each step and imitate the dancer's style with high fidelity through repetition and rehearsal. In both acts, we learn something about how we move, how another body moves, and how we perceive that movement—three phenomena that are fundamentally interlinked.

The act of recording our observations creates engagement that requires closer scrutiny, analysis, selection, and refinement of our experience. Activities like writing, drawing, painting, photographing, or filming our experience help save elements from the current moment to revisit them at another time and place—and even by another body. We can even view this practice of recording as a type of embodied, or physical, exercise. Not only does the mover have to engage in movement to record (as a finger depresses a camera shutter or as a hand holding a pen moves across the page, say), but also the recording becomes a basis of reflection.

We record our impressions of movement every time we take a photograph. Watching the subject in motion, we time when to capture the image based on their actions (often, this is as simple as waiting for the subject to become still with a smile, but sometimes it is more free-form than this convention). In this case, many of the choices made in capturing the impression of the moment are functions of the device's inner workings: the array of RGB (red, green, blue) values created by a digital camera is set by the device makers; the reflection of light on the device's sensors establishes what gets recorded into this format. On the other hand, Chapter 10 will introduce a movement notation system. However, we do not need complex technology or formalism to begin to translate the patterns we notice in movement to an archival format. A useful inroad is to record impressions in less structured formats, allowing us to discover more about the process of choice-making required in documenting our experience of movement.

For example, by using a pen and paper, we can watch a movement phenomenon and attempt to record what we see in real time. Forgoing any attempts to capture the details of the environment (i.e., features of the space where the movement happens) or the mover (i.e., height, clothing, or age), we can allow ourselves to focus on the changes in the scene, which can be roughly sketched through an ordinary writing implement. The examples in figure 1.4 show the results of such a process. These free-form sketches do not reveal any identifiable details about the observed event, and yet

Figure 1.4

Simple movement notation of four distinct observations of human movement. Here, time progresses left to right, and each line reflects impressions created by bodily movement over time. Each line break indicates a new body being observed in a new activity. The observation is of people working in a kitchen.

there is a sense of meaning: repeated loops suggest a repetitive movement; curved lines suggest softer, gentler actions than rigid, straight ones; and sharp changes in writing style suggest that the same may have been true in the observed movement.

The methodology is mostly known only to the notator—as is the value of the exercise, which is simply in *noticing* the choices made and the movement that inspired them more deeply. Rather than a perfect record of the event, this exercise is useful for observers to learn what they perceive. For example, here we could compare the same four observations by multiple observers watching the same event, noting similarities and differences in what is noted by each. This allows us to recognize other ways of observing the movement and, maybe most important, what the pattern is in our own observations. Just like a photograph or any other recording method, these scribbles are imperfect records of the event and represent (consciously or not) a value system in the choices made to create them. This is one of the central purposes of written symbolic representation (notation) of movement.

A second use of notation is to transfer movement from one body to another. Another exercise that we could perform with such scribbles is to share the record with a partner, asking them to move the script. This task could come with—or without—instructions of how the record was made.

For example, in the caption of figure 1.4, we are told quite a lot about how the observer used the paper to record impressions. But we might not have such instructions, so we could instead suppose that the impression is of a static pose, with one limb described by each line. Still, we would need to translate the design on the page into our body. And it will be instructive for the original observer to see how another person creates an interpretation.

As we add structure to our ability to describe movement, first through the taxonomy and associated symbology presented in part II and then through the notation system described in chapter 10, we hope that you will gain some bandwidth in how specific you can be in describing movement. To do so, you will add to your bank of terms for describing, and thus for noticing. However, this pure, free-form exercise of designing lines on a page amid live observation is one that can always be returned to for new discovery about, deeper kinesthetic attunement to, and richer noticing of any subject of interest.

Chapter Summary

In the act of making meaning with machines, we are wholly and fully consumed by thinking about how humans move, how machines move, and how, based on those respective models for movement, both entities interpret or perceive movement. We invite readers to explore this idea through creating a recording of their own kinesthetic attunement to some movement phenomenon in their own environments, as demonstrated in figure 1.4. This act will involve subjective choice and objective observation, producing a record that is both qualitative and quantitative and that reveals your own tendencies toward and styles of pattern making.

In reviewing prior scholarship from philosophy, dance, and engineering, we have opened this book by asserting that movement and first-person, subjective experience described through qualitative analysis are key to meaning-making; this book will deal with these topics with application to machines. Further, we have cast empirical measurement (often thought of as quantitative and objective) as a form of meaning-making, noting that when you set up an instrument to take a recording, you have made many subjective choices and need qualitative analysis to understand the result. Meaning relies on objective facts and quantitative measures as much as subjective perspective and qualitative description.[3]

2 Studying Movement: Somatics, Choreography, and Notation

Chapter 1 has set up the importance of the first-person subjective qualitative perspective in understanding movement, as well as how there are quantitative, objective, and measurable events that build to meaning in the eyes of human viewers. The goal of this chapter is to provide a background review of movement studies. The chapter will discuss three foundational parts of movement studies: **somatics, choreography,** and **notation**. While chapter 1 focused on a host of published academic studies, here we review some published work but also add more ethereal and personal outputs: performances, classes, and therapies. Thus, we use the term "movement studies" to comprise both movement *theory* and *practice*. To the former, principles and academic writing about somatics, choreography, and notation provide intellectual grounding for the specific taxonomies and principles employed in fields like dance, kinesiology, and physical therapy. To the latter, practical examples and anecdotes from embodied experiences in studios (see figure 2.1) and onstage situate this work inside a pragmatic, practice-oriented landscape for investigation and research.

2.1 Strategies from Somatics
Learning from the embodied, human experience

Somatics is a broad term used to describe movement methodologies that center around the experience of the human mover, inviting dualistic investigations into the mind-body connection through conscious physical practice honed through technique, principles, and repetition. The work of Irmgard Bartenieff, often termed "Bartenieff Fundamentals (BF)," is an example of such a method. Her work as a physical therapist and dancer

Figure 2.1
The setting for this chapter. Pictured here are examples of workshops in choreography, somatics, and notation, led by the authors alongside other movement analysts and movement artists from 2014 to 2020 at the University of Virginia (UVA) and the University of Illinois Urbana-Champaign (UIUC) inside the context of research in robotics. Unlike earlier workshops, which were held in a robotics lab in an engineering school (see figure 0.1 in the introduction), these workshops in dance department studios at UIUC had ample room for movement and incorporated special "sprung" flooring to safely support large weight shifts, including jumping.

helped her develop theories that supported a set of exercises used in her practice and meant as investigatory tools for others as well. Chapter 3 offers a longer discussion of this work, which is key to Laban/Bartenieff Movement Studies (LBMS) and the other information presented in this book.

Many other such **somatic practices** exist, including Alexander Technique, the Feldenkrais Method, as well as traditional forms of yoga and meditation. F. Matthias Alexander was a stage actor who developed a somatic method to improve physical performance; his work shares the approach of many somatic practices by encouraging the mover not to find "correct" patterns, but to notice habits and unconscious choices in order to find more possibilities. Likewise, Moshé Feldenkrais developed a physical practice to help refine, remove, or redefine habits. This work has been associated with alternative medicine and therapies. Thus, the form of these physical practices overlaps with the writings of Maurice Merleau-Ponty and Maxine Sheets-Johnstone, who were grappling with how to resolve the physical situation of their own bodies with their understanding of the world.

There is significant overlap and exchange among different traditions, and practitioners of one method are often trained in others as well. Moreover, credentials such as a Certification in Movement Analysis (CMA), Registered Dance/Movement Therapist (R-DMT), and Certified Feldenkrais Teacher

(CFT) attest to their holders' proficiency with aspects of the bodily knowledge represented in these practices. A professional association, the International Somatic Movement Education and Therapy Association (ISMETA), tracks expertise in somatics and offers its own credentialing. Much of the work in this community is passed down not through books or other writings (although those exist as well) but through physical, felt practice, as such, much of ISMETA credentialing is based on hours of experience and practice.

The authors of this book are both CMAs, and Cat is also a Registered Somatic Movement Educator (RSME) through ISMETA. However, the book reflects the totality of the authors' own background, which includes substantial training in dance more broadly. Much of this training occurred inside the paradigms of physical practice learned through studio-based, embodied transmission. This education is harder to track and cite, but we provide a few examples here (and in section 2.2) to give the reader a sense of the native environment of this type of knowledge. The following first-person excerpts describe somatic experience that transmitted physical embodiment inside a studio, led by movement experts.

Amy, in undergraduate studies at Princeton with Rebecca Lazier, 2006:

One task that might be set to a dancer is to roll from a supine position (lying on the floor on one's back) to a prone position (contacting the floor through the front side of one's body, such as kneeling on the shins). Babies do this action with relative ease but adult movers, like me at twentysomething, often develop difficulty in completing this action smoothly and efficiently. We'd been doing an exercise that involved such an action for a few weeks, as I remember, and Rebecca, in her characteristic exasperation, stopped us all to implore us to use our "head-tail" connection to ease this transition. I will never forget her rubbing her hand over the upper part of her spine, emphasizing the weight of her head at the end, a mechanical advantage, showing us the part of our bodies we were forgetting. Nor will I ever tire of the smooth feeling of rolling over my entire integrated spine to switch from supine to prone as I contracted my side and shifted my weight, supported by my heavy head, to roll over.

Cat, in undergraduate studies at Wesleyan with Mona Daleo, 1981:

We began class just recognizing our patterns of inhale/exhale. We then tried changing the ratio of the durations of the inhale/exhale, first creating longer inhales with short exhales, and vice versa, then balancing the duration of each. Then Mona directed us to use the inhale to expand and the exhale to contract. This was a modern dance class and up to then, after years of countless classes, I had never had breathing focused on as a "warm-up" to connect to my body. I was amazed to discover my inhale was much longer than my exhale—which, when exacerbated, created almost a "panting" that was limiting my movement capacity. As I linked my breath to movement—for example, by reaching my arms to the sides and then up over my head while inhaling and pressing my palms together to travel down the front surface of my body to rest in front of my chest on the exhale—I was stunned to discover a sense of my own inner space and contents as integral to my moving body; there was a sense that as the breath flowed in and out of my body, I could link that flow to support and clarify my movement execution.

We can never feel these experiences in the same way again. Our bodies have changed—we have learned new skills, we have aged, and new stimuli entreat us to move—but the knowledge created in those moments is more than a tacit, implicit coordination. It is a reasoned, articulated plan of action that comes from integrating sensations of one's own container into generalizable knowledge. When applying this knowledge to the design and analysis of a new tool (see, for example, section 4.5 of chapter 4), we say that we are employing a **somatic strategy.**

Such strategies, developed through the pursuit of full-fledged embodiment, have been present at pivotal moments of machine design. For example, in imagining how people could learn to use the power of computers for broad creative ends, designers at Xerox PARC utilized the metaphor of a "desktop," creating a digital space with a familiar embodied analog (Isaacson, 2014). At Apple, designers carve prototypes from foam and attach precise weights to simulate how the products will feel in consumers' hands (Isaacson, 2011). Such a practice better engages the body in the design process. Paul Dourish (2001, p. 126) defines such embodied interaction as "the creation, manipulation, and sharing of meaning through engaged interaction with artifacts." Zhang and Wakkary (2014) push further, highlighting how personal experience can and should influence industrial design.

Newer fields of human-machine design are emerging in robotics, the Internet of Things, and artificial intelligence. The field of human-robot interaction (HRI) inherits much from Dourish's and Zhang and Wakkary's approaches, merging design and phenomenology with engineering and science. Likewise, HRI is openly acknowledged as an interdisciplinary field. In a recent survey, Bartneck et al. (2020) describe the field as comprising engineering, science, and design, noting (as we do in this book) the difficulty in overcoming the barrier between working in science and engineering, where external, explicit models of knowledge prevail; and in design, where internal and implicit knowledge are often more celebrated. These researchers list many fields of contribution to HRI, although they do not mention any domains of the arts. However, they do acknowledge the challenge of embodiment and working with embodied agents as being a distinguishing aspect of the field, suggesting, in our view, that studio-based arts will have an increasingly important role in this field.

This studio approach that emphasizes *physical investigation situated through designers' own bodies* is a way to include somatic strategies in machine design, including in dance (Akerly, 2015). In this vein, the scholar Kristina Höök developed products with IKEA that featured lighted and heated elements designed for meditation and relaxation. She writes about her approach to product design as being explicitly informed by somatic practices, as well as the writings of phenomenologists. Emphasizing the importance of first-person experience, Höök promotes the need for physical attunement, embodied practice, and contextuality in design, eschewing the notion of "natural" interfaces for those that are designed with the full intelligence of the body, creating conventions that befit the context. Discussing Sheets-Johnstone's theory on the primacy of movement, Höök (2018, p. 35) describes the relationship between movement and meaning as well as the inextricability of our own movement experience from creating that meaning:

> As we move, meaning arises and is communicated; for example, you can see where I am heading by watching my gait and by knowing your own gait and movement. On that basis, it is easy to see how gestures, eye directions, or facial expressions can develop to become meaningful communication. It is with a basis in these prelinguistic, meaning-making practices that language can be invented and be filled with meaning. In that sense, language is post kinetic. It is through movement that we understand and act in the world. Our bodies move. Our thinking is movement.

This echoes Gillespie and Zittoun (2013)'s description of symbol formation in chapter 1: we create meaning through movement observed in context. Moreover, it adds an extremely important idea that we inherit from somatic practice: what we see in movement is based on our own containers, forms, and patterns for movement. And, as noted by Merleau-Ponty, our perceptions— our observations about the world—limit, shape, and form the meaning that we extract. Research in technology development has made similar points in the context of interviewing designers (Fdili Alaoui et al., 2015b).

When we think of the process of how people anthropomorphize machines, which is often explained as seeing things that have an inherent "human-like" shape moving in a "humanlike" way, we must note our own role in that. That is, we tend to anthropomorphize what we see *based on our own experience* because our human forms are our bases for knowing and making meaning in the world (Varella, 2018). This phenomenon is as much, if not more, about our own experience as any measurable feature of design or action we see on the machine.

Petra Gemeinboeck collaborates with choreographers and dancers who then embody simple, abstract shapes to kinesthetically probe the movement potential of nonhumanlike robot morphologies. This method, referred to as "Performative Body Mapping (PBM)," aims to "exploit one of the most interesting characteristics of robots from an embodied meaning-making perspective, that we can bodily resonate, kinesthetically extend into, and relationally make meaning with their spatial, embodied dynamics and the relations they spawn" (Gemeinboeck, 2021, p. 7). In this process, these researchers are including, in a direct way, nonanthropomorphic shapes (Gemeinboeck & Saunders, 2017), embodying the very real process of inter-pretation that other bodies go through when seeing unusual movement.

For example, as a dancer twists her elbow to enliven the corner of an abstract, three-dimensional form she is wearing, her body creates a map to meaning that gets filled in by the sensations of force, torque, and pres-sure on her musculoskeletal system and her own interpretations of sensing the resultant motion. While lay viewers will not know about this dancer's particular experience, they will find their own bodily maps to reinterpret the gesture. Perhaps it reminds them of a paper bag they saw blowing in the wind earlier in the day, or the time they got stuck pulling an intricate dress or shirt over their head. In either case, their prior and ongoing bodily experiences form their basis for reading the motion.

This example is leading from the somatic, internal perspective of a dancer to the external perception of the choices she makes—to her choreography. Acknowledging the embodied experience is an important advance that has created products that interact with our movement at increasingly high bandwidths and humanistic sensitivities (Höök et al., 2018). To consider the consequences of a machine that moves or otherwise exhibits agency in its environment, we must look to the crafting of movement in time and space—or *choreography*.

2.2 Tools (Technologies) from Choreography
Learning from the theory and practice of dance making

At first glance, "dance" as a concept can be seen as a prescribed list of movements that divides the world of movement between dance and nondance: either movers are dancing or they are not. Throughout this book, we utilize dance not as a set of allowable movements or instructions, but rather as a body of knowledge of **choreographic principles** for understanding, designing, and creating movement. The philosopher Catherine Elgin (2010, p. 81) describes the evolution of dance, across choreographer and genre, that crystalizes the process and body of knowledge that is choreography:

> *Swan Lake* is beautiful. It is delicate, graceful, enchanting. Martha Graham's *Night Journey* is not. It is riveting, harrowing, horrifying, often ugly. Yvonne Rainer's *Trio A* isn't even that. Being utterly pedestrian, it does not play on the emotions at all. But it is intriguing. Taken together these three dances raise questions: What is dance up to? What does it do and how does it do it? *Night Journey* discredits the thesis that the end of dance is beauty. *Trio A* discredits the thesis that the end is affective engagement. Possibly dance as such has no end. Different works and different genres pursue different ends. But whether or not dance has a telos, questions arise: "What does this particular dance do? How does it do it? And why?" My thesis is that dance embodies and conveys understanding.

Thus, dance (at least as we use the term) is not a list of movements, but rather the ongoing quest for new movements and new ways of expressing new ideas through movement. Dance equips us to break down, understand, and intentionally design movement (LaViers et al., 2018). Further, even within a given genre of codified dance, like ballet, flamenco, or break dancing, or of movement practice more broadly, like tai chi, gymnastics, or pole vaulting, there is a natural evolution of what its practitioners consider acceptable movements allowed by its teaching, practice, execution, and design as its cultural, scientific, and institutional setting changes. For example, the evolution from ballet to contemporary dance can be seen through a lineage of notable choreographers, such as Loie Fuller, Isadora Duncan, Katherine Dunham, Martha Graham, Doris Humphrey, José Limón, Merce Cunningham, Alvin Ailey, Yvonne Rainer, Bill T. Jones, Twyla Tharp, William Forsythe, Mark Morris, Ohad Naharin, and Elizabeth Streb.

Dance highlights the problem of trying to create a single movement "dictionary." Within dance, loosely speaking, there are many dictionaries:

one specific to each genre of dance, to each choreographer within a genre, and even to each dance within a given choreographer's body of work. Outside of dance, we will see many of the same movements used for very different purposes, with less formalization and codification of the taxonomies used. The wipe of a hand across the brow may be used in a particular Mark Morris piece, crafted with careful similarity across the myriad of bodies in his company to be performed "in unison"[1] and recalls his penchant for applying pedestrian-seeming gestures in precise relationship to complex classical music scores to convey a sense of humor. A similar wipe may be used inside a sweaty Southern church at the tragic funeral of a child; here, the gesture it is not used as an implement of artistic expression, but as a salve to soothe the internal needs of a particular mourner. A dictionary that claims to decode movement or enumerate a universal "body language," therefore, has limited intellectual merit and valid use cases because the number of such possible dictionaries is at least as large as the number of practicing choreographers (and probably closer to the number of distinct contexts and people using movement all over the world right now, in the past, or in the future). Moreover, "choreography," as we use the term, is a complex process, not just a lexicon.

The early postmodern choreographers of the Judson Church Group (1960–1964) used ideas from Jacques Derrida about deconstruction, rejecting prior forms and formalization, and in so doing offered a way to make meaning with movement that relied on the movement as being meaningful in itself rather than simply representing another idea. Merce Cunningham, with his randomization practices—for example, tossing a die before a performance to determine the order of movements just moments before the dance was performed—played with this notion too. While a work was created by deconstructing movement, separating prior forms into their gestural and postural component parts, a whole was in fact constructed that an audience experienced and interpreted as meaningful. This was a new way of creating dance and making meaning with and of movement.

As we grapple with defining movement and inviting the reader to engage in an embodied practice to support and inform construction of artificial embodiment, we give nod to this duality of deconstruction/construction and the use of the body as a site of knowledge and a tool for research (Cancienne & Snowber, 2003). In doing so, we hope to dispel the idea, which often permeates engineering and scientific communities, that choreography is useful only for making robots dance. As Elgin's writing (2010) helps explain, all movement can be dance and all movement can be not dance—it is the application, intention, and context that determine movement's purpose. Elgin's description of choreography as the pursuit of expression through novel movement that is not currently considered dance, along with the descriptions of the body as a site of information and choreography as a place of inquiry by Cancienne and Snowber (2003), reinforce this point of view. In other words, dance is not an isolated domain, and choreography is an intellectual pursuit for new knowledge that emphasizes the physical body.

As we will see in chapter 3, Rudolf Laban's work emerged from this viewpoint, formalizing principles of movement discovered in his work as a choreographer. Much of that work occurs through embodied transmission in a studio. Examples of such embodied absorption of the foundations of choreography include the following excerpts from our own studies.

Cat, in undergraduate studies at Wesleyan with Susan Foster, 1982:

> I took what was essentially a "dance history" class with Susan my junior year. She renamed this class "Anthropology of Dance" and taught us history from a perspective of choreographic taxonomies situated in different times and places. Typically, this type of class is taught from text, but we learned by trying on the movements of the choreographers from the different time periods we were studying. We also had to create movement compositions of our own using different choreographic techniques, or ways of making meaning, that the various choreographers we were studying had used. It had never occurred to me that different approaches to choreographing could create different meanings all together even when using the same set of movements. So, this was my first understanding of both the context-dependent nature of making meaning in movement, as well as the idea that making meaning from movement to express certain ideas can be accomplished in myriad ways that will all have an impact on both the performer and the audience (often in different ways).

Amy, in undergraduate studies at Princeton with Ze'eva Cohen, 2007:

> Ze'eva and I had a difficult relationship at times; something which cannot be separated from this embodied transmission (on the other hand, it is harder not to get along with the author of a textbook that you're reading). She (rightly) berated me for failing to account in my writing for the cultural conditions choreographers were working within. This "teaching moment," combined with my own contrarianism, cultivated an argumentative relationship between us. So in composition class, working with a fellow student, when Ze'eva was happy with our (in her words) "nice little dance" that was "almost finished," I felt inspired to push against the grain and shake up the entire piece. I suggested to my collaborator that we make a change to our work, which at that point used

contrasting movement profiles to represent successful, if difficultly achieved, communi-
cation. "What if I killed you at the end?" I asked her. In so many ways it was the polar
opposite of what Ze'eva expected from us, and it gave her the shock I'd hoped for. When
we showed the piece the next week, she gasped aloud. It was something I'd have never
tried on my own without the context Ze'eva and my collaborator created, which taught
me how countering the audience's expectations can work, how effective it can be at cre-
ating drama and communicating ideas. Instead of a piece about two beautiful women
learning to get along, we created a piece about female violence, dominance, and cunning.

Contemporaries of our teachers also explicated their knowledge in writing—often in nontraditional formats. For example, William Forsythe produced videos on DVD that explained his "improvisational technologies" (Haffner et al., 1999/2012) from which we derive our concept of **choreographic technologies.** This DVD set used visual annotations along with Forsythe's own nuanced embodiment (video of his movement was featured prominently in the work) to explain shapes, figures, and techniques for creating expression through the body (Haffner et al., 1999/2012). Additionally, Forsythe and his company collaborated with a team of technologists led by Norah Zuniga Shaw to document their embodied work in the piece *One Flat Thing, Reproduced* as a so-called choreographic resource (Delahunta & Shaw, 2006) in an online, interactive product entitled *Synchronous Objects* (Palazzi et al., 2009; Shaw, 2016).

Rebecca Lazier and the musician Dan Trueman wrote a book documenting her choreographic work, *There Might Be Others* (2016), where the onstage movement was created through live cues that allowed the dancers (and her, in real time) to make decisions about what emerged. The book also documented the process used to create the live improvisational musical score, as well as their collaboration with the engineering professor Naomi Leonard to design algorithmic movement strategies. Leonard also published work studying emergent behavior enabled by this collaboration (Özcimder et al., 2016, 2019). Such products document the knowledge produced through the process of choreography.

In addition to sharing knowledge from these domains for technologists' use, one of the themes of this book will be how working with machines helps develop and push the field of movement studies itself. Likewise, choreographers working with technology, like Margo Apostolos (1991), Thecla Schiphorst (Calvert et al., 1993), Stelarc (Atzori & Woolford, 1997), Susan Kozel (2007), Kate Sicchio (2014), Kate Ladenheim (Ladenheim & LaViers, 2021), and Catie Cuan (2021), are adding to this body of knowledge through their practice and corresponding theoretical developments. Lauren Bedal, a member of the Advanced Technology & Projects studio at Google, notes this phenomenon in an interview with *Dance Magazine*:

> Advances in gesture recognition technology require a new choreography of in-air hand gestures to interact with objects such as 'pinch' to select with Oculus Quest; 'swipe' to skip songs on Google's Pixel 4; and a 'bloom' to open the menu for Microsoft Hololens. Designers in this space are actively defining a new lexicon of movements to create intuitive and playful methods of interaction. (Skybetter, 2020)

Thus, choreography is already part of technology development, largely promoted through designers with niche, specialized interdisciplinary training. We propose that using choreography in a practical pursuit, such as developing machine behavior, is the implementation of choreographic technologies (e.g., see section 5.4 in chapter 5). We distinguish choreographic technologies from somatic strategies in terms of both the bodies of knowledge from which they derive and the impetus of the translation: choreographic technologies engage the environment with an external, expressive point of view, while somatic strategies focus on engagement with the body and germinate from a more internal, functional point of view. These distinctions can blur, however, given the relationship between internal sensations and our expression in the environment.[2]

2.3 Abstractions from Notational Systems
Formalizing many notions of movement

An early, robust form of recording movement comes from a surprising place: music notation. Here, viewing music-making as a subset of human movement activities, we investigate music notation to draw a parallel to it when we investigate notation for movement more broadly.[3]

The very first forms of music notation date back to 2000 BC (Schøyen Collection, 2022), when cuneiform tablets from the period have been found to contain a form of notation called "tablature" (Kelly, 2014, p. 18). This notation was specific to a particular instrument, (e.g., a lute). Rather than detailing notes and rhythms as modern notation does (notes would not be developed for another 3,000 years), this notation instructed the reader about how to play the instrument (e.g., pluck this string, then that string). We therefore say that this is a **platform-specific representation**: the instructions would not make sense when trying to perform the same patterns on a different instrument. Most modern programs for robots have this feature as well. That is, the same program, or set of movement instructions, cannot be used interchangeably with most robots.[4]

Another way of thinking of this early notation is that humans did not yet have an abstract understanding of music: it was a concrete thing that emerged from a particular tool when you acted on it in a particular way. In the Middle Ages, people would become curious about the prospect of "[learning] songs which they had never heard" (Kelly, 2014, p. 65). To do so required abstraction—centrally, the invention of the music *note*, which prescribed a sound *pitch*. Thus, this abstraction enabled the development of **platform-invariant representation,** which allowed the same song (or series of pitches) to be played on different instruments and sung by people with different vocal qualities and ranges.

The musical note was identified by a monk, Guido Monaco, around 1030. This advance changed earlier forms of notation that came after tablature, which used *neumes* to preference pitch over timbre, quality, and melodic form. As Kelly (2014, pp. 79–80) writes:

> It was true before Guido, and it remains true after him, that not every aspect of a performance, not everything about a piece of music, gets transferred to the page. You have to choose what's important. When Guido invented the note, as it were, he threw out a lot of other information.

Thus, even the process of notating music, a commonplace and widely used technique, is an act of *subjective* choice and established *convention*. This system of notation has become the basis for the digitization of music, automatic music players, transmission and archiving of songs, and other advances. All of this is enabled by a *particular* abstract understanding of the design space of music. Each distinct form of understanding may be thought of as different **notational abstractions.** Other methods, such as shape note music notation, continue to advance the representation of music with additional convenience, detail, and applicability (Marrocco, 1964; Wong & Danesi, 2015).

Such audible vocalizations and felt rhythms are a subset of what we might try to notate in bodily movement more broadly. That is, the design space for music is much, much lower-dimensional than the space of movement options for the *whole* body. Why was this subset the first to be notated? In Guido's time, music played a central role in religious processions and services, and there were few devices to aid the process of archiving and preserving the structure of these religious implements. However, with the dawn of photography, recording movement from an external, quantitative lens became possible before any such comprehensive movement notation akin to that for music was developed. Namely, notes were invented over 800 years before sound recording (Thomas Edison's phonograph), and that was followed closely by video recording a little over a decade later (Edison's kinetograph), which allowed movement instances to be recorded easily. This access to recording devices may have thwarted progress in the broader development of movement notation, leaving the abstract ideas that were required for distinct bodies to perform the same actions relatively unexplicated, embedded in human minds and bodies. There is an urgent contemporary need for understanding movement abstraction.

One pioneer in pushing the commercial use case for this idea is JaQuel Knight, a prolific choreographer of some of the most famous dance routines of popular music artists, including Beyoncé's acclaimed *Single Ladies* music video. While the music that accompanies this work of art is well protected by copyright and intellectual property law, Knight's much-repeated and riffed choreography is not. He was paid a single fee for his work on the video, while the director, songwriters, and singers are entitled to royalties for reusage of the work. Why do these artists get compensated differently from Knight? Because they have the ability to represent their work: the video itself can be saved as a media file, the words of the song are written in language, and the music of the work is notated. Knight, by contrast, faces an uphill battle to say what movement constitutes his choreography (Milzoff, 2020).

It is no surprise, then, that Knight has turned to movement notation to begin to document and copyright his work. Knight has worked with Lynne Weber at the Dance Notation Bureau to use movement notation to file for intellectual protection of his choreography. This notation is much like the early form of music notation, tablature, in that it does not lend itself well to abstraction across bodies. How should a quadruped robot perform his choreography for *Single Ladies*? How should a wheelchair user perform it? These questions are similar to those that tablature writers might have had about how to perform the same song on different instruments, and they are especially pertinent when thinking about alien bodies—for example, machines—creating meaningful movements.

Some attempts have been made to use extant notational schemes for movement to encode, prescribe, and interpret motion for machines (Barakova et al., 2015). Naoko Abe and Jean-Paul Laumond led such an effort, collecting the work of many researchers (Laumond & Abe, 2016) and conducting their own work (Abe et al., 2017) in this vein. However, most of these attempts fall victim to the deterministic interpretation of anthropomorphization, assuming that using a robot of the correct morphology allows transmission between human and machine.

For example, in work with a large, hulking "humanoid" (with no malleability in the core), steps encoded in movement notation were transcribed to the robot (Salaris et al., 2017). Since many of the distal joints, which the system of notation they used focuses on prescribing, are to some extent mimicked in the machine, there is a case to be made that the team succeeded. However, the robot is a wildly different mechanical device than the original human performers (and observers), with different ranges of motion and dynamic access. Thus, the re-creation can be thought of as a cartoonlike imitation that does little to unpack the assumptions made in creating the translation. Which part of the machine maps to which part of the movement score is a subjective choice: for example, they reflect assumptions about the human form that include two arms in the upper body and two legs in the lower body.

We lack explicit and consistently used abstractions for notating gross bodily movement and, as such, the translation process is especially brittle; it breaks down for many machines and human bodies that do not share the same distal morphology as able-bodied humans, entirely missing the expressive and functional nature of our core anatomy. Yet there are examples of success: for example, green blobs can be highly effective bodies for re-creating the same movement ideas (Kingston & Egerstedt, 2011). Sara Hendren (2020), in grappling with the intersection between bodies and technology, explains the opportunity presented by diverse bodies, which she views as generative sites of creativity and inspiration for designing a better built world. Likewise, removing our ableist assumptions about bodies can pave the way for deeper understanding about what bodily movement (*perceived* change) is.

Chapter Summary

This chapter has introduced somatic practice and choreography as key bodies of knowledge comprising movement studies. It has also begun to provide fodder for how these foundational bodies of knowledge will be used to construct and deconstruct movement in applied settings as *somatic strate-gies* and *choreographic technologies*. We also introduce the notion that the act of notation explicitly brings the choices of human observers into the process of recording movement. Thus, the problems of both generating and recognizing the "same" motion across bodies require a system that offers a process for uncovering the myriad instances of human experience of move-ment, as well as offering design tools for generating observable patterns in that movement. Chapter 3 introduces such a system, which involves a par-ticular set of somatic strategies, choreographic technologies, and notational abstractions.

3 Constructing Movement: Somatic Strategies, Choreographic Technologies, and Notational Abstractions through a Laban/Bartenieff Lens

What are the options in a mover's palette? As we saw in chapter 2, there are many ways of answering this question—contemporary and historical, formal and informal, qualitative and quantitative—and, in our view, all of them fall short. In short, the answer to this question is unknown. New feats of athletic prowess are accomplished; new dance moves are invented; and new technological devices extend our own bodies. This book is centered around one way of describing a mover's palette, using a particular system that we will expand and refine in our presentation. As has been set up in the previous chapters, we care about three aspects of this system: its ability to capture both internal, somatic knowledge and external, choreographic knowledge, as well as its inclusion of a process for notating movement based on capturing essential patterns. This chapter will provide historical context and a contemporary overview of this system and introduce the embodied way of working that is essential to understanding it. Then, in part II of this book, we provide expansive coverage of this body of knowledge that integrates ideas from multiple practices, as well as our work with machines, to provide a primer for movement studies.

We frame the options in a mover's palette as aspects of both their own interpretation and their own ability to move and distinguish elements that come from their own experience (somatics) from those that come from communicating with others (choreography). We describe applying such distilled knowledge to the development of technology as the use of a somatic strategy and choreographic technology, respectively. A "strategy" is a plan of action designed to achieve a major or overall aim, and "embodied" refers to expression in concrete, often corporeal, form. In this book, a "somatic strategy" is an intelligent action taken by a physical body that achieves a goal. A "technology" is the use of knowledge for practical applications,

and "choreographic" relates to the expertise and knowledge used by choreographers in expressing ideas through performance with moving bodies,
creating an event that is meaningful for the human participants. In this
book, therefore, a "choreographic technology" is a practical application
of choices made about articulated bodies moving in space and time that
considers context and audience for a particular expressive end. Notational
abstractions developed from this body-based practice of movement are crucial to consider in this text as well, as such systems become the basis of
transfer of movement between bodies. An "abstraction" refers to dealing in
ideas rather than events, and "notational" refers to the practice of symbolic
representation. In this book, consequently, "notational abstraction" refers
to the concept that our representation of movement depends on our ideas
about it. We are interested in the implications that this work has for designers, researchers, and technologists—namely, by posing the question: how
can a given movement idea be perceived as the "same" when performed by
distinct bodies?

3.1 Historical Overview
Development of movement studies in the Laban/Bartenieff tradition

This section provides a picture of **Laban/Bartenieff Movement Studies
(LBMS),**[1] which has grown from **Laban Movement Analysis (LMA), Bartenieff Fundamentals (BF),** and the contribution of many practitioners who
have worked with the material over the past 100 years or so. We begin
with a description of the system that has emerged from LBMS as it exists
today. We then give a very brief chronology of how the system developed,
culminating in the diverse arenas where the work has been applied. Finally,
we provide some context about how this book presents extensions to the
system that have come from our own work with machines.

3.1.1 Contemporary Snapshot
Training programs in movement analysis[2] do not occur in lecture halls; instead
they reside in dance studios, both inside and outside academic institutions.
Thus, the work is often not written down—it is often not even articulated
in language at all. It is executed live, in person, in the body, with rhythms
and gestures, through experiences crafted by experts in choreography and
presentation through movement. Transmission often occurs moving body to

moving body and is disseminated by teachers to students who then go on to utilize, adapt, and craft experiences of their own, which are in turn transmitted to their students—and so on—without being tied to written prose.

This embodiment of the work necessitates an explicit statement about the bodies who originated this work, their historical context, and the limited swath of people who have studied the work since. As Rudolf Laban writes in *Principles of Dance and Movement Notation* (1956), as quoted by Irmgard Bartenieff in *Body Movement: Coping with the Environment* (1980/2013, p. 17):

> We should be able to do every imaginable movement and then select those which seem to be the most suitable and desirable for our own nature. These can be found only by each individual himself. For this reason, practice of the free use of kinetic and dynamic possibilities is of the greatest advantage. We should be acquainted both with the general movement capacities of a healthy body and mind and with the specific restrictions and capacities resulting from the individual and structure of our own bodies and minds.

This quotation first highlights the goal of movement studies: to offer choices to a mover. Further, it points to pitfalls that occur in evaluating such choices when many such evaluations are unrealized and unconscious. Likewise, the history of the development of LBMS contains specific moments that reveal those often ugly unconscious biases.

3.1.2 Historical Development

As a choreographer, Rudolf Laban was interested in human movement, and he approached his work primarily from the perspective of dance and mysticism. He developed a system to "write" movement (Guest, 2013, p. 11) and "map" the space of the mover (Studd & Cox, 2020, p. 139) called **Labanotation**. This work also established broader ideas about the relationship of the mover's space to their expression and relationship to the larger world (Laban & Ullmann, 1950/2011), resulting in what has become known as "space harmony" (Dörr, 2007). Later, he became interested in dynamic aspects of movement and began collaborating with an expert in time-motion studies, F. C. Lawrence (Bradley, 2008). This area of inquiry was further developed and refined into what has now become known as the "Effort component" of movement analysis (Laban & Lawrence, 1959). This work, in whole, became known as Laban Movement Analysis (LMA).

Irmgard Bartenieff was a physical therapist, educator, and dancer who used her training in Laban's work to pioneer the use of dance in a therapeutic context. While working with diverse populations, including professional ballet dancers and polio patients, she developed principles that helped her patients find for themselves new (and/or renewed) patterns of motion. For example, both the ballet dancers and polio patients with whom she worked had developed patterns of muscular holding in their pelvises and lumbar spines that prevented coordination between their upper and lower bodies. Thus, she furthered the system by bringing the physical body explicitly into the framework. Through these efforts, she developed her own set of movement methods and exercises that are often referred to as "Bartenieff Fundamentals (BF)." This evolution has continued with today's practitioners further developing Bartenieff's original principles and applying this work to a variety of applications, including hands-on physical therapy, diagnostics for injury, improvement of connectivity within the body, repatterning to avoid reinjury, and improved physical crafts in dance, sports, public speaking, and other activities requiring physical presence and performance. This work is considered an important extension of Laban's original advances in the space of choreography that helps facilitate physical expression by a variety of bodies. This work has become an important, integrated part of the system overall, and thus, although LMA is a better-known acronym to describe this community of work, many prefer the term "Laban/Bartenieff Movement Studies (LBMS)" to acknowledge Bartenieff's contribution.

The system continued to evolve after Laban and Bartenieff, as others contributed to clarifying, explicating, and naming parts of the movement experience. Many of these insights were integrated into four components of LBMS—Body, Effort, Space, and Shape (BESS)—but people also created other systems based on their interests and applications of the material. Notable in this list are Warren Lamb, who developed Action Profiling, seeding what later became its own component of the system, which describes the shape and relationship between posture and gesture (Lamb & Watson, 1979); Judith Kestenberg, whose work with infants and mothers led to the creation of Kestenberg Movement Profiling; Martha Davis, who studied nonverbal communication (Davis, 1983); Bonnie Bainbridge Cohen, who developed the well-known Body-Mind Centering (BMC) somatic practice (Cohen et al., 1993/2012); Peggy Hackney, who created new symbologies and methodologies for connecting breath to body (Hackney, 2003); Colleen Wahl, who added clarity about phrasing's significance in all aspects of the system (Wahl, 2019); and Karen Studd and Laura Cox, who have codified and evolved a new modern taxonomy (Studd & Cox, 2013/2020).

Laban's early work began in Germany and partly flourished during the rise of fascism. While he later fled Nazi Germany and ended his days in England, this period working under the Nazi regime was his most prolific and well funded (Bradley, 2008). Moreover, his work was created within the context of Western culture and indeed privileged that culture's art—and the white, able bodies that it exalted—above others. Alan Lomax later used Laban's work to value one way of moving (styles of movement used in Western art) over another (styles of movement used in art from indigenous communities), claiming that certain styles of movement were more advanced, and likewise were the cultures that used them (Williams, 2007). In contemporary times, similar analysis gets applied to political candidates' movement patterns and actions, which often serves to justify the observer's own political leanings. Such applications attempt to measure movement, assigning value to particular choices, instead of valuing all movement options. Moreover, acknowledging the influences of the observer's own bias is essential to moving the work beyond white supremacy, anti-Semitism, colonialism, and patriarchy.

To that end, LBMS has been moving forward, as seen in many arenas where this embodied taxonomy and system for describing movement (Maletic, 2011) have been used to advance human knowledge in a variety of intellectual domains. The system is used today by a heterogeneous group of people working across multiple disciplines, all of which include a component of human movement, including but not limited to health care, the creative and performing arts, education, sports, diplomacy, and conflict resolution. Some notable examples include the following:

- **Creative and performing arts:** Cadence Whittier uses LBMS to examine and enhance ballet pedagogy, including in *Creative Ballet Teaching: Technique and Artistry for the 21st-Century Ballet Dancer,* her 2017 book that explores using LBMS in ballet technique. Dancers working in the theater have also explored exercises based on LBMS to enhance actor training in both the vocal and physical realm to enhance the expressivity of body and voice (Bloom et al., 2017).

- **Social work and social justice:** Martha Eddy (2016) uses LBMS and somatic practices to explore embodied peace education, as well as violence prevention. Her research seeks to understand and explicate the role of the body in negotiation. Deborah Heifetz utilizes LBMS to teach international conflict resolution and mediation in culture and community development. She has served as special advisor to the Crisis Management Team of the Israeli police and she supports peacebuilding by utilizing embodiment practices (Fernandes, 2014).

- **Sports and biomechanics:** Many people in the sports and rehabilitation fields use LBMS to enhance athletic performance and help rehabilitate movement (Hamburg, 1995). Janet Hamburg was a leader in this field and is best known for using LBMS to develop a comprehensive seated exercise program for people with Parkinson's disease (Hamburg, 2004). Mark Morris's "Dance for PD Program," led by David Leventhal (McRae et al., 2018), uses Hamburg's work.

- **Psychology and therapy:** Rachelle Tsachor and Tal Shafir (2019) use LBMS in their research to enhance understanding of the connection between body movement and emotion. This work is teasing out the fundamental underpinnings of applied fields like Dance/Movement Therapy (DMT), in which practitioners work with patients in a therapeutic context through embodied and creative movement.

- **Movement theory and technology:** Many people connecting technology to movement have found LBMS to be an apt resource for thinking about movement in a structured manner. For example, in developing forms of movement representation for technological development and interaction, Nicolás Salazar Sutil (2015) studies Laban's work extensively, examining connections to contemporaries in psychology, like Jacques Lacan, and in philosophy, like Maurice Merleau-Ponty.

- **Affective computing and dance:** Thecla Schiphorst has led several studies that examine how computational tools can enhance dance practice, several of which employ LBMS. In one work, sonification is explored as a method to improve dance pedagogy (Françoise et al., 2014); in another, her team examines how experts see movement with greater nuance than those without specialized training (Fdili Alaoui et al., 2015a); and in another, tools for choreographers are developed (Fdili Alaoui et al., 2014).

- **Artificial intelligence:** The computer scientist James Wang is working to model human expression based on bodily movement by creating an annotated data set, the Body Language Dataset (BoLD). They use LMA to support their characterization of expression and attempts to automatically recognize bodily expression (Luo et al., 2020). LMA has long been thought of by scientists and engineers as a resource for "decoding body language." However, LMA is not a lexicon for "what movements mean"; instead, it offers a system for resolving salient actions and inactions that is informed by context and the observer's own preferences and bias. Wang and his team are working with produced media (e.g., movies and YouTube videos) from a US-centric view. Such works share expressive strategies—ones that are often not replicated in natural, nonperformative behavior (Gladwell, 2019)—that LMA is useful for describing, allowing the team's algorithm to match raw data and lay impressions from human workers on Amazon Mechanical Turk via LMA-based annotations.

- **Animation:** Just as it is a useful tool for choreographing dances, LMA has been used to design the actions of virtual avatars and animated characters (Badler et al., 2000; Durupinar et al., 2016; Ziegelmaier et al., 2020; Kim et al., 2022). Leslie Bishko (2014), an animator and CMA on the faculty of Emily Carr University, focuses on expression through movement in animation. She uses LMA in her research to support movement visualization in three-dimensional animation, including puppeteering and creating avatars in virtual environments.

- **Robotics and automation:** Likewise, a growing number of roboticists have incorporated the use of movement studies in **expressive robotics**, particularly those working in human-robot interaction (HRI). For example, Heather Knight and Reid Simmons (2014) use Laban's theory of effort, which we describe in chapter 8, to develop movement profiles for robots that observers associate with motivations in the form of emotive states (happy and sad), personality traits (shy and confident), and situational states (rushed and lackadaisical). Similar work has been done in numerous research groups, employing both Laban's theories (Barakova & Lourens, 2010; LaViers & Egerstedt, 2012; Kim et al., 2012; Sharma et al., 2013; Burton et al., 2016; Cui et al., 2019; Bacula & LaViers, 2021) and Bartenieff's later work (Nakata, 2001; Huzaifa et al., 2016). Although associations with machines are not explicitly part of LBMS, these researchers use the system to structure their ideas about movement and then validate the associations for a particular context through empirical studies with human subjects. These research efforts suggest the importance of LBMS for HRI.

- **Biology and animal behavior:** The Effort component has also been used to refine observations of animal movement (Gladwell, 2009). For example, Robert Fagan et al. (1997) used the system to examine locomotion, social, and even manipulation behaviors in gorillas, canines, goats, and magpies.

Within the LBMS community, Cat, alongside Karen Studd, Laura Cox, and Esther Geiger, has been refining, evolving, and developing the system in both content and pedagogy through ongoing international training programs, working under WholeMovement, as well as with certification programs like the one that Amy completed as a student. The influence of their collective approach is therefore formative for the material presented in this book. In particular, this group is using a taxonomy that includes identifying new basic body actions (e.g., vocalizing); fundamental baselines of the body experience (i.e., weight-sensing and flow-sensing); patterns of body organization that align with motor development (e.g., radial symmetry); recognizing and articulating the Shape component as having primary patterns of change of toward and away from self (i.e., gathering/scattering and concave/convex); and creating the attendant symbols to further clarify notation and its role in understanding movement (e.g., modifying the "forward" and "back" space symbols) (Studd & Cox, 2013/2020). We continue to evolve and refine the system in this book, adding and refining symbols, editing the presentation of topics, developing new approaches to notation, and delineating a new fifth component (Time)—all emerging from our work with machines. The rich and collaborative practice in the community of movement analysts keeps the system vibrant and relevant to multiple fields of application and strengthens the system itself.

3.1.3 How Working with Machines Has Evolved the Material

This book comes most directly from our work and movement experiences with (and as) engineers. We employ LBMS and other theoretical constructs in a way that has proved useful for our field of research. In particular, we add a new component, refine the presentation of certain symbols, and edit concepts for consistency, especially with an eye for translating ideas across natural and artificial bodies. We also end the tradition of associating Laban's and Bartenieff's names with the work, opting for a generic acronym to name the system.

Our evolution repeats an established pattern. In working with patients with differing movement capacities, Bartenieff had to explicate many things about the body—ways of describing it, ways of patterning it—that she would not have found if limited to her own bodily experience. Likewise, studying the movement of babies has been useful for understanding the system: adult movers come with many mature patterns of movement that younger, immature movers have yet to develop. This practice is common to other movement theories like those of Moshé Feldenkrais and F. Matthias Alexander, and one of the most popular videos shared across these communities is of a young baby rolling over for one of the first times, through a series of exploratory kicks and weight shifts. This video allows us to see foundational patterns in movement that build to complex, three-dimensional, cross-lateral adult human behaviors. Thus, working with bodies different than one's own has a rich and generative history in somatics as well as choreography (examples of choreographers working with machines are given in section 2.2 of chapter 2).

Likewise, in working with machines and algorithmic descriptions of movement, we have found bodies with novel capacities distinct from our own. While even the most advanced robots do not possess the same intelligence (or embodiment) of adult human movers, machines can be used to create starkly pure action—for example, a movement that is strikingly to the left, void of any sagittal or vertical contributions that exist in a human gesture. This raises new questions about how to describe movement.

The description of LBMS in the next section is a particular instantiation of the system that we find useful for working with machines, and as such, it is an extension of the history described here. This extension is informed by our own particular perspectives, which are centered in Western, proscenium-based forms of movement expression and conditioned by growing up and working in the US in the context of robotics and movement analysis. As a result, in this book we are using and expanding the material in our own applications, as limited by our own cultural and intellectual embeddings.

3.2   Laban/Bartenieff Movement Studies: The BESST System
Presenting a unified framework for movement analysis of natural and artificial bodies

There are many systems that address choreography and/or somatic practice, as described in chapter 2, but we value the lens of LBMS, as we find that it bridges these two approaches to observing, creating, and experiencing

movement. In other words, both the inner experience and outward manifestation of movement—of both the individual and the individual in relationship to the environment and other movers—are explicated. LBMS does
not favor one approach over the other; rather, it constantly engages in an
ongoing duality of personal and universal, inner and outer, self and other,
and provides explicit connections between the mover and the environment.

Take, for example, ballet technique. It is a specific form of expression,
preferencing a particular vocabulary that aims to extend movement potential beyond the "normal" range, characterized by external rotation, extreme
flexion and extension, and extraordinary agility, strength, speed, balance,
and finesse. All of these aims establish a clear sense of technical expertise,
and such virtuosity is measured against an external form relative to an
idealized execution. As a result, there is a set of expressions in ballet technique that are judged to be virtuosic, and this privileges an external presentation of self as the larger goal.

Conversely, a somatic practice like BMC preferences the inner or experienced movement of the practitioner and is focused on virtuosity as an
achievement of inner connectedness, efficiency of movement, and support of breath—in other words, the experience, sensations, and feelings
of the mover. In BMC, practitioners explicate connection to internal body
systems, including organs, and sensory experience; this favors an inner
approach of self-awareness as the larger goal.

In LBMS, virtuosity is achieved through broadening the palette of movement options in both sensing and action. The goal is not to execute a specific
vocabulary of movement (as in ballet), but rather to broaden the range of
choices that a mover has. The goal is not about honing a particular inner
experience (as in BMC), but rather to develop the ability to make clear
choices in movement manifest in the environment. We see LBMS as a process
with which to engage our bodies to continue to learn, rather than a fully
formed, prescriptive paradigm. Moreover, as the human experience continues to broaden, such as through the use of new tools like robots, this body of
knowledge will continue to evolve to accommodate that. Box 3.1 presents the
BESST System, comprised of five components with overarching themes (each
with an associated symbol, shown in figure 3.1) that we use for these ends.[3]

The acronym "BESST" is created through the first letter of each component
(Body, Effort, Space, Shape, and Time). The five themes form a description of
recurrent patterns of relationship between seemingly opposing or distinct

Box 3.1
The BESST System

- **The Body component:** Addresses "what" is moving. The body is the basis of our physical and lived experience. (See chapter 4 for more about this component.)
- **The Space component:** Addresses "where" movement is happening. Movement occurs in our internal, personal, and shared environments. (See chapter 5 for more about this component.)
- **The Time component:** Addresses "when" movement happens. Movement occurs over time, giving access to ideas like duration, phrasing, and rhythm. (See chapter 6 for more about this component.)
- **The Shape component:** Addresses "for whom" the movement is happening. This captures the interaction of the body in space, naming higher-order patterns that emerge from this phenomena, such as those of accommodation of and relationship to the environment. The larger idea here is the interaction of our body in space and the bridge of inner to outer and self to the larger environment. (See chapter 7 for more about this component.)
- **The Effort component:** Addresses "how" the movement is happening. This captures the behavior of the body in time, naming higher-order patterns that emerge from this phenomena, such as those that relate to distinction in movement quality. The larger idea here is recognizing the motivation and intent that drives complex movement expression. (See chapter 8 for more about this component.)

In identifying and reconciling patterns, paradoxes emerge. This notion is incorporated in the system through the concept of "duality,"[4] visualized through a lemniscate (or Möbius strip)—a form that expresses two seemingly distinct surfaces actually blending together and moving seamlessly from one to the other. As you trace the inner edge of the surface, it becomes the outer edge and back again, in an endless loop that is at once two surfaces and one form. This physical form is used to describe the overarching themes of the system:

- **Stability/Mobility theme:** Addresses the interrelationship of two primary movement descriptors, acknowledging that to be a stable moving body, one needs access to movement, or mobilization, and to have access to movement, one needs stability. (See section 4.6 in chapter 4 for more about this theme.)
- **Function/Expression theme:** Addresses the interrelationship between practical and aesthetic dimensions, acknowledging that to be efficient, one needs a broad palette of options, and to be expressive, one needs functional support. (See section 5.5 in chapter 5 for more about this theme.)

(*continued*)

Box 3.1 (continued)

- **Exertion/Recuperation theme:** Addresses the interrelationship between work and relaxation, acknowledging that work requires periods of recuperation, and recuperation becomes exertive without periods of work. (See section 6.5 of chapter 6 for more about this theme.)

- **Self/Other theme:** Addresses the relationship between our personal experience and the broader world, acknowledging that we are a world of connected individuals. (See section 7.5 of chapter 7 for more about this theme.)

- **Inner/Outer theme:** Addresses the relationship between the inner intent and experience and the outer expression and perception of a mover, acknowledging that one's outer expression changes one's internal state and vice versa. This theme is also directly reflected in the topology of the lemniscate where the outer surface becomes inner and vice versa. (See section 8.6 in chapter 8 for more about this theme.)

ideas within the system. Additional dualities relevant to movement studies and engineering are seen in other common pairings (e.g., yin and yang) that may sometimes be presented as a linear spectrum with opposing poles but here are understood as two versions of the same idea; that is, as "two sides of the same coin." Consider the following list for other possible dualities:

- *Explaining methods of analysis*: Quantitative/qualitative, subjective/objective, meaning/measurement, micro/macro, individual/universal, simple/complex, part/whole

  - For example, a whole is made up of parts; each part is its own whole.

- *Explaining design*: Form/function, construction/deconstruction

  - For example, form creates function; function dictates form.

- *Explaining movement*: Condensing/expanding, content/container, constant/change, success/failure

  - For example, when one surface of the body is expanding, another surface is always contracting (and vice versa).

- *Explaining agents*: Mind/body, sense/act, intelligence/embodiment, mechanization/computation

  - For example, mechanization creates computation, and we use computation to generate mechanization.

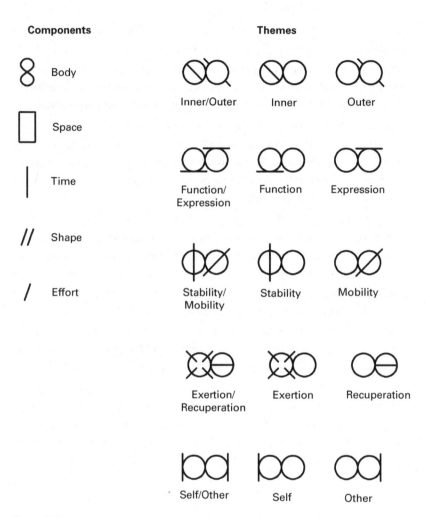

Figure 3.1
Component and theme symbols. The component symbols establish a visual pattern as they are used in many of the symbols for concepts within that component. For example, the same slash used here for Effort is found at the center of the Effort graph in figure 8.2 in chapter 8. The theme symbols also set up a feature of many other symbols in the system; they demonstrate how related concepts are represented with symbols that can aggregate and overlay one another. For example, the symbols for Function and Expression, which can be used on their own, also can be superimposed to create the symbol for the Function/Expression theme.

The relationships between components in the BESST System are called **affinities.** For example, activities that we often do above our heads share associations, just as activities done close to the ground do. These activities, happening in similar areas of the environment, inherently share features of movement. For example, extended arms and a sense of destabilization are associated with reaching overhead due to our bodily forms, while crouched lower limbs and a sense of connection to the earth (or grounding) are associated with squatting low to the ground. Our experiences create associations between these activities, linked by common geometry and dynamic consequences of similar bodies.

We visualize these affinities in two ways: one that is nonhierarchical and one that is hierarchical. In the nonhierarchical model, these components are not seen as independent features of movement, but instead as a series of interrelated, overlapping, and redundant ways of describing parallel perspectives of movement. This model articulates how in mature adult movers, each component of the system relates equally to all the others. An adult mover can generate movement from our bodies (foregrounding the Body component) or from our inner motivation (foregrounding the Effort component). In the hierarchical model, we recognize the foundational roles of body, space, and time in enabling the movement of both natural and artificial bodies—and thus give primacy to the Body, Space, and Time components. This model recognizes that all bodies can move, and as that movement comes into increasingly complex relationship with the environment, expression—in space, in time, with other bodies, and governed by the mover's own intent—becomes more and more evident. These two models are illustrated in figure 3.2.

In the nonhierarchical model of relationship between components, the components are viewed as equal. In the hierarchical model of relationship between components, we note that we can perceive a body (whether natural or artificial) moving in space over time, but when meaningful choices are made regarding changing our form to accommodate and/or interact with the environment, or changing our expression to broadcast our intention, that is the manifestation of a mature mover/observer (e.g., an adult human). We can assign these things to robots by projecting our own experience (as we do, say, in seeing expression in cartoons) and by recognizing that robots are bodies that move in space and time, so they can reveal shape and expression.

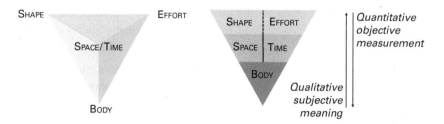

Figure 3.2
Nonhierarchical and hierarchical models of the BESST System components. The non-hierarchical model uses a tetrahedron (left) and the hierarchical model uses a triangle (right). The former model follows the traditional presentation of BESS, where each component is seen as a possible central basis for analyzing movement. The latter model arranges components in an ordered manner that reflects the types of movement concepts that robots (or other simple bodies) can more feasibly embody.

Thus, in this model, Shape is defined as patterns (of form change) in space, and Effort as patterns (of expressive/dynamic change) in time.

In analyzing the movement of machines, we observe richer patterns in the adult mover (natural) and simpler expression in the machine mover (artificial). Yet both of these are bodies that move in space and time. We see this reflected in an often-stated observation in the LBMS community: "Body is basis." Therefore, in this book we point to the primacy of the Body component, followed by Space and Time, which round out a foundational picture of a body moving. As we consider the movement in context, especially that of the mover's environment, higher-order concepts about the mover can be revealed, shading the baseline established through a body moving in space and time with ideas about relationship (via the Shape component) and intent (via the Effort component).

This taxonomy also enables symbolic, visual representation, a notation system called **motif**, which allows us to capture the essential elements of movement events, which in turn can allow us to both perceive larger patterns and overall meaning. The symbols are designed to combine very specific ideas of movement into a larger whole through visual representation. This also allows us to find new patterns and practices, as we can start from motif and then find movement, and/or we can look at movement and then write motif. Chapter 10 introduces several systems of notation and details several ways to use motif.

Thus, we begin to explicate the idea that movement studies offer a palette of movement choices that in turn, coupled with the ability to make symbolic representation, can support observation and—through both experience and perception—make meaning, find patterns, and engage bodies fully with their knowing, moving selves. That conscious awareness forms a new basis for making design choices regarding the movement of machines and how we can in turn make meaning of machines in motion.

3.3 Using Embodiment as a Process of Inquiry
Developing the physical practice necessary to examine and experience embodiment

Humans are always using movement, both perceived and experienced, to make meaning of our world. We are pattern makers and pattern perceivers, and it is through this constant process that we strive to reconcile paradoxes. Insofar as we "know" anything about movement, this knowledge is constructed inside the containers of our body: *we know because we move*, and *how we move determines what we know*. The BESST System is a way of languaging and experiencing that constructed, inherently personal reality by naming the parts of movement that coalesce to create meaning.

So far, we hope that we have set up a sense of how many people working with many forms of scholarship have grappled with understanding the perceived components of patterns in motion. We hope that you have a sense of how complex and unfinished this conceptual space is—and realize that the tools we offer in this book have grown out of a limited, Western point of view that is structurally and pedagogically distinct from traditional technical disciplines. Most important, we need to communicate the value of moving in your own body to understand this work. The physical knowledge of moving one's body contains a wealth of experiential, qualitative, and subjective information. Though complex and unfinished, this space is concrete, and it is where an emerging articulation of sensational, visceral, embodied, and personal understanding is situated.

Why do you need to move your own body to learn this material? Can't everything be written? In writing about the development of music notation, particularly the switch to notes, which prescribe pitch, over the more qualitative *neumes*, which were used in the further past to indicate emphasis and melodic shape, Kelly (2014, p. 80) writes that "you can't write down everything. Whenever you choose one thing, you inevitably omit some-thing else." As notation brings a fuller conscious attention and awareness to specific aspects of the process of moving and interpreting movement, it can enhance aspects of that lived experience—but possibly at the expense of ignoring other aspects. Thus, a personal, lived, *embodied* practice (and repetition of practice) are needed to integrate the concepts we introduce in part II. With this in mind, we pause here to scaffold this suggested practice.

We invite—nay, implore—the reader to participate in the embodied exer-cises given at the end of key sections of this book, including here. We call these exercises "embodied" to emphasize their physical, corporeal nature, but we could just as easily call them "intelligent," as they are designed to help hone the reader's capacity for understanding. Just as intelligence is a feature of human capacity, we note that embodiment is a similarly impor-tant feature that means almost the same thing but foregrounds the physi-cal, lived experience rather than the mental, generalized one. Here, we aim to offer a series of prompts to guide you in developing a somatic practice that you can carry throughout this book.

Embodied Exercises

- **Getting moving and tuning in:** The goal of this exercise is to begin to engage in and learn from a somatic experience, which will require *think-ing, feeling,* and *sensing* your movement.

  ◦ Start by finding a comfortable standing position. Take a few moments to be intentional about the width of your feet, the balance of your weight, the alignment of your spine, the tension in your arms. Try to find a comfortable, not overly held or overly relaxed posture. In com-ing to stillness here, notice what you feel in your body. Namely, tune in to all the movement that continues even here, in stillness.

  ◦ Place your hands on your chest to notice the rise and fall of your breath. Place your hands on your heart to feel your pulse. Is there pain or sensation anywhere that draws your attention? Touch that part of your body and rub or tap it. See if that changes the sensation.

  ◦ Notice your stance. How is the weight of your body distributed through your feet? More weight on one foot? Are your feet aligned under your legs or spread out? What is your resting stance?

- Now begin walking around the room. Notice the room. Notice yourself moving within the room. What draws your attention?

- Continue moving around the room while increasing and decreasing the speed of your walk, escalating into a run and cutting through the space. Then freeze. Notice your feet contacting the ground. Has your sense of your stance changed? Take a look. What do you see? What do you sense? Once again, place your hands on your chest. Has the rise and fall of your breath changed?

- Place your hands on your heart. Has your pulse quickened? What parts of your body are now drawing attention?

Try repeating this process several times, using the following exercises to revisit it through new lenses, noticing each time what changes and how your awareness of your body changes.

- **Synthesizing somatic experience:** Try to synthesize the previous exercise. Consider and welcome both the positive and negative aspects of your experience. Looking through the lenses of thinking, feeling, and sensing is one route to doing this.

  - What did you think about this exercise? Did you gain any new knowledge? Was it familiar or unfamiliar?

  - What did you sense during this exercise? Did you knock into anyone or anything? Did you trip and/or fall?

  - How did this exercise make you feel? What was comfortable or uncomfortable about this exercise? Did you enjoy any specific aspect of it? How did it change the way you feel?

- ▣◀ **Introduction to the interrelated lenses of the BESST System:** Take a video, repeating the event in the "getting moving and tuning in" exercise, for easier viewing and analysis. Then, respond to each prompt, introducing the components of movement studies discussed in the next five chapters.

  - Body: *What* are you sensing? Which parts of your body are involved in the action?

  - Space: *Where* are you moving? What are you thinking about and attending to?

  - Time: *When* is the movement happening? What happens first, second, third, and so on?

- ○ Shape: What or *whom* are you interacting with?
- ○ Effort: What were you feeling during the movement? What does it look like you were feeling? What is the inner motivation for this action? *How* was that manifested in the way you performed the motion?

- **Gait matching:** Here, you will explore the process of kinesthetic attunement through basic locomotion. If you do not have full use of your legs, translate the task through your own method; while moving from bipedal walking to, say, rolling a wheelchair is a big translation, all individuals will have a mismatch between their bodies and the bodies that they try to imitate. Here, we are interested in more abstract concepts like order, pacing, rhythm, muscular tension, posture, movement quality, and attitude.

  - ○ Fall in step behind someone (perhaps along a corridor in your building, keeping a safe and respectful distance).
  - ○ First, match the order and pace of their legs, moving the left and right sides of your body in step with theirs. Then, look deeper: what kind of weight shift are they using? How are their arms moving? What is the posture of their spine, and how does it change across the gait cycle?
  - ○ Now reflect on how this imitation feels. Is it comfortable? Empowering? Awkward? Which body parts feel as though they are in alien territory? Which feel right at home? How does this inform your relationship (or how you would imagine yourself to be engaged in relationship or activity) with this person?
  - ○ Return to your own natural gait, shaking off the act of watching another person. How does your own body feel now? Are you newly noticing any aspects of your own gait?
  - ○ 📹 Record yourself inhabiting different gaits and come back a day or two later, when the experience is no longer fresh, to offer yourself an external point of view of your own movement and how (and if) you perceptibly varied it to accomplish the exercise.

Chapter Summary

This chapter has introduced the main influence on the movement studies presented in this book: LBMS. We have worked to provide a brief historical context for how the BESST System came to be. The chapter has also introduced the notion of embodied exercises. These activities allow you to engage in your own movement, recognizing your interactions with others, and illuminating the idea that you are already using movement analysis in your daily life. The goal of this book is to help you articulate that expertise and bring it to the point of applying it to technology design through somatic strategies, choreographic technologies, and notational abstractions.

Part I has set up some subtleties of systematically viewing moving bodies—a phenomenon that we are constantly surrounded by and experiencing ourselves—and outlined a path forward. Next, in part II, each chapter will deal with Body, Space, Time, Shape, and Effort in turn. That said, while being useful for the format of a book, such a linear presentation is in tension with the holistic, overlapping, and redundant nature in how these categories are used (particularly in the nonhierarchical model shown at left in figure 3.2). Each component will introduce a set of symbols, utilized to deepen the specificity of ideas that can be notated with motif. In part III, we bring in case studies from technology, introducing systems of notation and highlighting open problems in the field.

II Describing Movement with an Embodied Taxonomy: The BESST System

4   What Is Moving? The Interconnection of Body Parts and Action (Body)

Part I of this book introduced the complexities of noticing, studying, and constructing movement, valuing both the subjective experience and the objective event, which combine to create a meaningful pattern. This helps us to notice and describe a moving phenomenon and then make new choreographic choices or interpretations about it as two separate, interacting processes. This chapter will zoom in on one aspect of this picture: the physically moving body. At times, it will seem like an anatomy chapter, but we are concerned much less with the true, scientific anatomy of the body than the *perceived* anatomy. It is through the perception of our own bodies that we are able to craft movement behavior that gets us through the tasks of our day, including communicating and interacting with others around us; it is through our perception of others' bodies that we can work together toward the ends of accomplishing such tasks. In this frame, we present the *Body component* of the BESST System to heighten our understanding of moving bodies.

It is also important to notice how different bodies represent a unique and valuable perspective on the world. We see this idea quite literally in the viewing of any artwork, including in the performing arts. Imagine the distinct, physical perspective that a person who is five feet tall has on a painting like the *Mona Lisa,* hung with its base five feet from the floor, versus someone who is seven feet tall. They will see slightly different shapes on the canvas due to the projection of the painting at their respective heights: the shorter person has to actually look up to view the thirty-inch-high canvas, while the taller person must look down to see it. The two viewers will have different experiences navigating to the painting; the taller person can see the work over the heads of other onlookers as she approaches, while the shorter one does not see its features until reaching the front of the viewing line. We know that we will have readers who move through the world differently than we do. Whether it be above-average height, an old injury, the use of a wheelchair, uncommon speed, or any number of features that shape and define human bodies, these attributes contribute important and novel access to the environment, shaping unique experiences. Thus, rather than emphasizing a singular view of embodied experience, this chapter aims to equip individual readers to investigate their own bodily experiences and physical intelligence, which, in turn, inform how they view the movement of others.

Machines also have bodies, but their mechanical architectures and their programmed behaviors do not align with the patterning of human bodies (a fact that is often clouded by the naming of robot components and subsystems as "arms" and "legs", for example). What we are interested in is the way these bodies move in space and time. By first developing a richer taxonomy to describe the patterns of our own container, we can better compare and contrast our actions with machines' actions. In fact, distinct body architectures (both human and nonhuman) successfully express the same or similar ideas in many contexts and applications. For example, in this chapter, we will see a low-dimensional (simple) robot that successfully expresses distinct styles of walking gaits in a user study. Even though this simulated body does not have "knees"—or even exist in three dimensions—patterning the artificial body in a particular way evokes ideas of "skittering," "sauntering," and other gait styles. This example shows how studying the body through a somatic lens can broaden the types of perceptually salient behaviors that robots can perform.

Thus, this chapter explicates features of our conscious experience of our own container (and the contents of that container). It explicates fundamentals, reviews anatomy, and provides physical landmarks from which to reference motion. We will also begin to introduce some associated symbols related to these ideas that can be used both to abstract and to notate larger movement ideas.

4.1 The Body as a Container
How our anatomy affects our experience of and ability to accommodate the environment

The body is a container—not only of our lived experience, but also of bones, muscles, organs, tissues, and fluids. Our perception of movement (and the broader world) is affected by context and our own individual characteristics, including the relative shapes, sizes, and compositions of our bodies. Take, for example, the two authors of this book—both are women, white, petite, and have similar experiences, particularly regarding our dance and movement training. However, we are of two different generations (we are twenty-five years apart in age) with limbs of different lengths, ranges of motion, and strength.

Given our commonalities, we share many physical biases; given our differences, we show how even distinct bodies can move in unison, imitating one another via abstraction. For example, a person in a wheelchair has yet another experience of the body as a container, but whether on wheels or on feet, the bipedal moving body and the wheelchairing body both *locomote*. Both bodies can move forward in space from one location to another. The meaning of that locomotion will be understood differently but can in many ways be identified as "the same." The meaning and experience of the movement will always be affected by context, including the body creating and observing it.

Another example of this is height. A tall, sighted performance goer can move through a large crowd, scan the assembled bodies, and see a large field of movement, while a short, sighted performance goer cannot. These two bodies have different physical forms and thus afford different experiences and assimilated intelligences about the world. The short person will see those in front of and around him, but will not be able to see the crowd. This engenders a different experience and creates the need for different strategies for accommodating an environment. For example, in the context of viewing a performance in a theater, a shorter performance goer is more likely to sacrifice a broad view of the whole stage and select a seat in the front, magnifying the detail of the bodies onstage and muting the broader picture, providing a distinct experience (quite literally) of the viewed performance. After the performance, a performance goer seated up close might comment on the sweat of the performers that belied effort that was otherwise well hidden, while a performance goer seated farther away might comment on the affecting distance between the protagonists during a soulful duet. A blind performance goer may have picked up more acutely a particular quality of contrast between the timbres of the performers' voices.

If the performance goers were octopuses, they would have even more different ideas about the world. This kind of thought experiment becomes useful in developing machines because we can (and do) create eight-legged machines that may need to coexist in human-facing settings. What is common between the human form and an octopus? Even without expertise in human or octopus biology, we can list a few rough, obvious commonalities, which form a basis of shared experience and potential understanding, noting whether motion occurs at the core or in distal elements, the relationship to gravity, the body's midline, and preferred directions for action and sensing. These concepts are defined in box 4.1 and given symbolic representations in figure 4.1.

In this book, we present the list in box 4.1 (and figure 4.1) as a basis for the mechanisms of anthropomorphism. Puppeteers and animators explicate

---

Box 4.1

Features of Expressive Bodies

- **Core versus distal motion:** One way to look at moving bodies is to assess and assign centrally located moving elements as opposed to distal/edge elements. Often, the **core** houses ongoing life processes in animals (e.g., heartbeat and breathing) and critical functions in machines (e.g., clock processors and cooling fans), while distal body parts that are considered **limb** deal with more direct interaction with the environment, making this division particularly meaningful in many contexts.

- **Relationship to gravity:** A consistent relationship to gravity is a defining factor in experiencing a body, as all of the bodies considered in this book interact inside a gravity field that immediately affords a "top" versus a "bottom"—or an "axis of length," as discussed more thoroughly in section 4.4—for each moving agent.

- **Midline:** Bilateral symmetry is a common feature of complex, active bodies (defined as bodies that exhibit complex interactions with the environment, e.g., through manipulation, and displayed in members of the chordate and cephalopod phyla) and establishes a sense of "right" and "left." This sense can be imagined as the line of symmetry of our bilateral forms; this is often termed the "midline" of the body. In dance, the midline is often used to reference the notion of having one side and the other side. "Crossing the midline" occurs when a body part breaks across this imaginary boundary.

- **Preferred directions for action and sensing:** We experience the world as bodies that have a preferential direction for locomotion, establishing a clear notion of "front" and "back" (on the body) or "forward" and "backward" (in motion). This establishment of up/down, left/right, and front/back, immediately suggests a location for a "face," which may be understood both as a central location of sensory organs and a set of movement features that do not severely affect the stability of the body. They can consequently be used to express something about the internal state in a wide variety of physical engagements (e.g., with arms full, locomoting across a gravelly environment, etc.).[1]

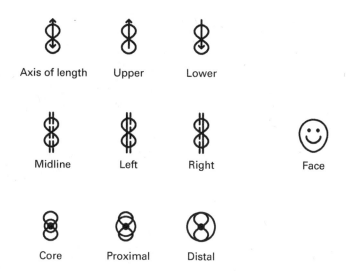

Figure 4.1

Symbols for features of expressive bodies. These symbols correspond to the concepts listed in box 4.1. Symbols for core, proximal, and distal help distinguish between core versus distal motion. A symbol for axis of length, which can be modified to indicate upper versus lower, represents a body's relationship to gravity. A symbol for midline, which can be modified to indicate the right versus left side, represents a body's bilateral symmetry. Face is a general symbol meant to represent a body's preferred sensory center, which in humans is also a highly expressive bodily feature.

similar principles in their process of animating inanimate forms (even simple strips of foam) into lively characters with personality that can express a narrative story, and this list describes features common to many (if not all) fictional cartoon characters (Thomas et al., 1995). Examples of this vast array of bodies, all of which can be expressive for human viewers at varying levels of abstraction—from simple, cartoon figures to detailed, real bodies— are illustrated in figure 4.2.

Several specific features are common to many (but not all) humans and are worth noting and naming explicitly as well. There are: geometric relationships that commonly arise in dealing with the human form, including **patterns of body organization;** bony landmarks, often palpable areas of the body (where bone protrusions can be felt through the skin) that form waypoints in experiencing our form; connections between bony landmarks called **kinetic chains;** muscular landmarks, often referred to as "diaphragms" (a thin sheet of material forming a partition), which delineate key areas of

Figure 4.2
Examples of bodies and simplified, yet still expressive correlates. Even though each
cartoon (bottom row) is simpler than the original body that it represents (top row),
it obeys the principles listed in box 4.1.

action; and aggregate anatomical terminology aimed at describing our *perceived* anatomy (rather than detailing it formally).

The lists in box 4.2 are just a start, as any geometric relationship, muscular pattern, and bony landmark can be used to facilitate a sense of connection based on the goal of the mover. However, these terms have become
commonly used in movement studies because they arise frequently in
physical practice that focuses on the gross shape change of the whole body
(e.g., dance, yoga, and rehabilitation). Likewise, a symbol set of body parts
and patterns of body organization that are commonly referred to is given
in figure 4.3.

Embodied Exercises

• **Finding bony landmarks:** Go through the list of bony landmarks and see
  if you can feel each one through palpation and weight shifts. Pay special
  attention to the complex structure of each joint. For example, the pelvis
  is not of a homogeneous weight, nor is it a static "bowl"; it allows subtle
  shifts of bone to accommodate weight.

Box 4.2
Features Common to Many Human Forms

- **Geometric relationships:**
  - *Six limbs*: We count the two arms, two legs, head (the upper part of the spine), and tail (the lower part of the spine), to create a starfish image that captures our ability to locomote and have complex interactions with the environment. (Even for bodies missing an arm or leg, this image captures the access to movement afforded through the proximal region of the missing limb.)
  - *Patterns of body organization*: These relationships result from our six-limbedness and our preference for two of them (head and tail).
    - **Radial symmetry:** A primal pattern that recognizes six limbs (head, tail, arms, legs), as well as internal content (such as bones, muscles, and viscera), as equal, either moving in toward the self or out toward the environment as one whole (like a jellyfish ambiently floating or swimming).
    - **Spinal:** A primal pattern that emphasizes the spine as the body's core, with the movement relationship between the two ends of the spine a central organizing feature that supports the throu–ghline of our form, particularly in our length (like a fish swimming).
    - **Core/distal:** Recognizes all limbs as equal, moving toward or away from the core (like a hanging pull-string wooden puppet toy whose arms and legs move relative to the stable core of the puppet when the string is pulled).
    - **Head/tail:** Recognizes the primacy of the spine from head to tail in organizing movement, emphasizing the weighted ends of the head and pelvis (like a cheetah running, using the head and tail in oscillation to propel its action).
    - **Upper/lower halves:** Recognizes the upper unit (e.g., arms and head as one) as separate and equal to the oneness of the lower unit (e.g., tail and legs), like a frog hopping.
    - **Right/left halves:** Recognizes the whole right side as one unit and the whole left side as another unit and defines a clear sense of midline and bilateral symmetry, like a lizard bending left or right.
    - **Cross-lateral:** Recognizes the connection of the upper to the lower from right to left through the core and crossing the midline, like an upright human walking with their arms swinging gently in opposition to the legs.

(continued)

Box 4.2 (continued)

- **Bony landmarks:**
  - Spine
    - Cervical (7 vertebrae, C1–C7, with C1 being closest to the skull)
    - Thoracic (12 vertebrae, T1–T12)
    - Lumbar (5 vertebrae, L1–L5)
    - Sacrum (5 fused vertebrae)
    - Coccyx (4 fused vertebrae)
  - Pelvic girdle
    - Sit bones
    - Pubic bone
    - Tailbone
    - Sacrum
    - Iliac crests
  - Hands
    - Wrists
    - Heels of the hands
    - Balls of the hands
    - Fingertips
  - Feet
    - Ankles
    - Heels of the feet
    - Balls of the feet
    - Toes
- **Kinetic chains:**
  - Head/tail
  - Heels of the feet/tail
  - Scapula/fingertips
  - Scapula/head
  - Scapula/sternum
  - Hand/other hand
  - Head/heels of the hands
  - Heels of the feet/sit bones
  - Sit bones/sternum

(continued)

Box 4.2 (continued)

- ◦ Head/hands
- ◦ Hand/scapula
- ◦ Trochanter/other trochanter
- **Muscular landmarks:**
  - ◦ *The throat (partition between the upper core and the head)*: This esophageal muscle group, near the occipital joint at the peak of the spine, creates the basis for vocalization.
  - ◦ *The diaphragm (partition between the upper and lower body)*: This layer of muscle bisects the human form around the T12 vertebrae and is the key mechanism of breath, pressing down on an inhale (creating a larger volume of relatively lower pressure around the lungs) and pressing up on an exhale.
  - ◦ *The pelvic floor (partition between the lower limbs and the core)*: The pelvic floor is a group of muscles and the associated connective tissue that interweave to form support for the organs of the pelvis and for shifting weight near the center of mass (located between the pubic bone and the tail bone in the bowl of the pelvis).
  - ◦ *The psoas (connection between the lower and upper body)*: The iliopsoas muscle begins at T12 and spirals through the pelvis to insert at the head of the femur (thigh bone). It is one of the deepest muscles, and primary hip flexors, that connects our upper and lower halves and is critical to our ability to stand upright and locomote efficiently.
  - ◦ *The palms of the hands (partition between the environment and the body, frequently)*: The surface of the skin on the underside of the hand that runs between the tips of the five fingers to the wrist; it provides a surface for object manipulation, gesture, and weight bearing.
  - ◦ *The soles of the feet (partition between the floor and the body, frequently)*: The surface of skin on the bottom of the feet that runs from the tips of the toes to the heel and is what is in contact with the floor (both sensing and applying forces) when we walk.
- **Aggregate anatomical terminology:**
  - ◦ *Core*: Also referred to as the "torso," the core is generally recognized as the area between both shoulders and both hip joints. It includes the spine, with the crucial meeting place of T12–L1 (the 12th thoracic vertebrae and the 1st lumbar vertebrae, where the diaphragm is attached and further delineates two volumes) marking the link of upper core to lower core—also referred to as the **center of levity** (approximately the

(*continued*)

> **Box 4.2 (continued)**
>
> centroid of the skull, collarbone, sternum, and associated abdominal muscles and vital organs) and **center of gravity** (approximately the centroid of the pelvic girdle, iliac crest, sacrum, and associated abdominal muscles and vital organs). The skull and pelvic girdle are often described as the 5th and 6th limbs.
>
> ○ *Proximal joints*: Forming the basis of the "limbs," the proximal joints are considered to be the shoulders and hips, but if the head and tail are experienced as limbs, then the sacroiliac joint (where the bottom of the spine meets the pelvis) and occipital joint (where the top of the spine meets the skull) are also proximal.[2]
>
> ○ *Distal joints*: The distal joints are considered to be joints associated with the hands and feet.[3]
>
> ○ *Prone/supine*: Refer to two different surfaces of the body, often clarifying the relationship to the ground (prone is belly and face down, and supine is belly and face up).

- **Heel rock:** This exercise is analogous to a frequency response test in systems analysis, where unknown systems are probed with sinusoidal inputs to characterize their features. We will similarly use an oscillatory input to the body to explore, experience, and edify its connections. This exercise can be practiced in many positions and is classically experienced while lying supine on the ground with extended limbs. To begin, we recommend trying it in "supine hook," lying supine with the knees bent, allowing for full contact with the floor through the bottoms of the feet. Then move on to the other "neutral" postures suggested in section 4.2.

  ○ Lie supine on the floor, grounding your connection, especially through the heels of your feet; this forms a "neutral" starting state.

  ○ Notice each point of contact that your body has with the floor: heels, pelvic girdle, chest girdle, back of head, arms.

  ○ Pushing into the connection of your feet to the ground, give yourself a "rock," pushing into this point, increasing the reaction force with the floor, and then release, relaxing back to "neutral."

  ○ Repeat, playing with different rhythms, arrhythms, and tempos. Notice the connection between bodily parts—in particular, how the

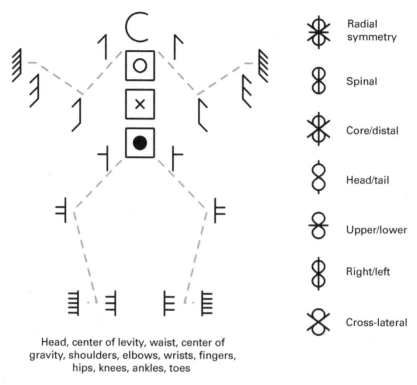

Radial symmetry

Spinal

Core/distal

Head/tail

Upper/lower

Right/left

Cross-lateral

Head, center of levity, waist, center of
gravity, shoulders, elbows, wrists, fingers,
hips, knees, ankles, toes

Figure 4.3
Symbols for body parts and patterns of body organization. The symbols for body
parts (left) are arranged along a stick figure to roughly show their anatomical positioning on the human form. The symbols for patterns of body organization (right)
leverage the Body component symbol to abstract the human form.

push you give with your heels ripples through to your head. How does
that feel? Is there anywhere that ripple gets stuck? Use the bony connections listed previously, like heel/tail and sacrum/scapula, and as
each part contacts the floor, pay attention to what order and in what
timing these sensations occur, to notice this experience in more detail.

*It is a learned skill to increase the sensations and physical actions that
your body allows in this exercise. As adults, we develop all sorts of patterns
of "stiffness" and "holding" in our bodies. We resist the relaxed undulation
that this exercise is promoting, especially when it is unfamiliar. You can return
to this foundational exercise many times throughout the book, using it as a*

preparation for other exercises, revisiting it with breath and other bodily sensa-
tions. Always work to increase your ability to notice and experience this gentle
rocking and the successive action of each body part rocking into another and
transmitting that pulse from the heel up to the head—and, paradoxically, you
will need to do less to facilitate that increased motion.

4.2   Finding a Neutral Baseline
Investigation with weight-sensing and flow-sensing

The body is constantly in motion. Maintaining any given position or state
requires sustained action guided by sensory feedback loops. Even what feels
like "sitting still" is not actual stillness: the human body is constantly filled
with micro-movements and -adjustments. Dancers and performance artists, like mimes, train for years to achieve the appearance of total stillness;
yet, for example, they cannot eliminate the beat of the heart that is constantly filling the body with subtle reverberations. So, then, if everything
is in constant motion, how is it possible to recognize that a new coherent
movement, such as a particular gesture, posture, or action, has occurred?

One way that such changes become salient is in contrast to a neutral point
of reference—or **baseline**. It is hard to find a true, universal neutral baseline
because this idea is contextual, personal, and adaptive: the answer differs
between a ballet and tai chi class and is different for people of different flex-
ibilities or with different numbers of limbs. A different baseline should be
established for a study of movement of office workers sitting at their desks
than one used for studying basketball players shooting free throws. There
are several body positions that are common choices for finding and establishing a neutral baseline in the LBMS tradition: lying on the ground in an
I position, an *X* position, supine hook, and standing with parallel legs. These
positions are described in box 4.3.

There is a perceived experience of our body, which these neutral positions
help us observe and is distinct from its physical truth. For example, we do
not perceive the exact weight (as measured in pounds or kilograms) of our
body or its parts; however, we do have a sense of *having* weight, an idea that
is affected by the mood, emotion, preoccupations, and goals of the moment.
Lying in a familiar posture on the floor helps us check in with how our
weight feels *today, right now.* It is difficult to estimate exactly how fast or with
what acceleration any one of our body parts is moving at a given moment,

---

Box 4.3
Body Positions for Establishing a Neutral Baseline

- **Lying on the ground in an *I*:** Lying supine with limbs extended down (the form typically used for Savasana in yogic traditions).

- **Lying in an *X*:** Lying supine with the arms overhead and away from the body at approximately 45 degrees, with the legs extended away from the body in the same manner.

- **Supine hook:** Lying supine, the arms by the side, with the soles of the feet on the floor, and the knees bent and pointing toward the ceiling. The legs are parallel to each other and the heels are in line with the sit bones.

- **Standing:** Standing with the arms relaxed by the side and the feet under the hips roughly parallel to one another.

Each of these will offer different experiences depending on whether the body is passively or actively engaged (or somewhere in between).

---

but our sense of its relative motion may be described in some moments as free and ongoing or more rigid and controlled in other moments.

To describe these ideas about experience, the Body component of the BESST System distinguishes broader, perceptual ideas of **weight** and **flow** that are related to but distinct from the quantities of the same names that are measured by scales and meters. There is an entwined relationship between these ideas of weight (both active and passive) and flow (both release and control). Flow foregrounds the connection to the environment and others in it; weight foregrounds the connection to our body/container. Practically, these ideas are very useful in generating different styles of movement and intention within movement through greater attunement to our own bodily state.

Weight (as differentiated from the expression of "weight effort" described in chapter 8) emerges from our experience of our mass in relationship with gravity. This is the process of **weight-sensing**. This sensation allows us to differentiate ourselves from the world, noticing the "other" and finding "me," as opposed to "you" and/or the environment. The ability to activate our agency through the sensation of our weight gives us the ability to act separately from others. Flow (as differentiated from the expression of "flow effort" described in chapter 8) arises from our experience of our contents in ongoing motion and from connecting to the ongoingness of our

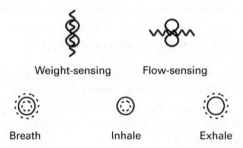

Weight-sensing   Flow-sensing

Breath   Inhale   Exhale

Figure 4.4
Symbols for weight-sensing, flow-sensing, and breath. Top: symbols for weight-sensing and flow-sensing[4] use the Body component symbol and a wiggly line (less sharply undulating than the vibratory phrasing bow in figure 6.1 in chapter 6) along the weight and flow effort axes in the Effort graph (shown in figure 8.2 in chapter 8). Bottom: symbols for breath,[5] including inhale and exhale.

environment. For example, ongoing processes like heartbeat and breath[6] can be viewed most broadly as modulations of control and release. Noticing and partaking in this process is called **flow-sensing.**[7] Symbols for these concepts are shown in figure 4.4.

Another way of understanding weight- and flow-sensing is to draw parallels to the ideas of "doing" and "being," respectively. When we activate our weight, there is a consciousness of doing something, but when we connect to our flow, there is a consciousness of being something. In this way, weight-sensing[8] begins to allow an experience of differentiation between self and environment, and flow-sensing facilitates an experience of undifferentiation and connection to the broader environment. It is from this experience of weight and flow that we come into awareness of our body and its neutral, baseline state, from which we can perceive distinct movements occurring. As our movement becomes differentiated from such a baseline, we make qualitative and dynamic movement choices to express that intent.

As we find ourselves in increasingly frequent and close relationship with machines, we want to predict and be able to "know" the state of the machine. If we perceive its "readiness" and "presence" from subtle shifts in posture that simulate these ideas of weight- and flow-sensing identified in human movers, we may better understand the machine as part of our environment. Thus, we endeavor to find a way to make machines indicate readiness. Such an idea is exhibited in software updates to the SoftBank

NAO robot, which now features an Autonomous Life setting that introduces subtle, seemingly random shifts and built-in "animacy" that simu-
lates these ideas in the machine. Seeing the action of the robot gives people
an ongoing signal that the machine is on and available for function. In a
similar way, we signal readiness to other humans by returning to our baseline. The experience of baseline weight and flow, which is implicit most
of the time in humans, therefore becomes critical to explicate in order for
machines to express that same engagement to us.

Embodied Exercises

The following sequence of movement experiences is designed to take the
mover from a place of "being" to a place of "doing"—in other words, from
a state of undifferentiation from the whole of the environment to one of
differentiation between individual agency and the environment. We call
this weight-sensing/flow-sensing in the BESST System.

- **Breathing exercise to connect with flow:** Begin by lying on the floor and
  allowing your weight to release into the floor and your breath to flow
  freely and easily through your body. As you inhale, imagine the breath
  flowing like water inside your body and bathing your inner space. As
  you exhale, release the flow into the environment. Allow the inhale/
  exhale phrase to move easily and evenly.

- **Sloshing of internals to connect with flow:** Begin to allow this flowing
  breath to move you—pouring your breath and sloshing your internal
  contents to allow different parts of your body to come in contact with
  the floor. As different surfaces come in contact with the floor, use that
  sensation to begin to identify the container of your body.

- **Bouncing to connect with weight:** Begin to activate pushing against the
  floor and shifting your weight to a level change (i.e., sitting, kneeling,
  and standing), and finding different places of support for your weight.
  As you do this, add movements of pulsing, jiggling, and bouncing to
  find your weight, activating it in relationship to gravity.

- **Mobilizing in space through level change and locomotion and checking in
  on contacts with floor to connect weight:** Clarify your movement intention now to include changing levels, going to different places in the room,
  making different pathways, and using different forms of locomotion (e.g.,

rolling, crawling, and walking) to experience agency. Continue to notice your relationship to the floor and the different surfaces of your body.

- **Repeat:** Start the process from the beginning to find more connections, sensations, and experiences that increase nuance and range of expression.

4.3   Basic Body Actions
Naming changes in the container

Now that we have established useful patterns in the form of the body—largely reviewing selected elements of established anatomy, albeit through a lens for embodiment and somatic experience—and clarified a notion of baseline from which changes may be more salient, we can move to establishing a taxonomy for describing patterns in the body's motion.

In this section, we align with a commonly recognized idea in computer vision, robotics, and animation: the movement primitive discussed in section 1.2 of chapter 1. This idea has been posed to extract coherent movements from two-dimensional video and three-dimensional motion capture—the so-called *motion segmentation* problem. While it is not clear what quantitative formulation will become dominant for this purpose, the plethora of papers pursuing this analysis confirms what experience suggests: there are moments in time that make sense to chop up movements, and other moments that do not. Listing out types of possible movements is one step—which has been used by engineers as well—that helps better formulate this problem. But how do we choose these types? How much do these choices bias our resulting analysis? This is a step where added nuance in sensation, observation, and intuition provides formal guides and tools for explaining results—both in individual experience and in the output of computer algorithms.

At its core, any movement is a shift in the center of mass of the body. Changing the position of a hand—outstretching a palm toward an object in the environment, say, or even just fidgeting slightly as one sits alone waiting for a bus—changes how gravity is acting on a body. Whether this outstretched hand represents the intent to pick up an object or this fidget indicates nervousness or relaxation is a result of inference about objective changes in the body container. Thus, we would like a set of terms that allow us to describe *without assigning a meaning* to such changes in center of mass or **change of support.** To this end, we suggest a list of movement concepts that offer more descriptive power, called **basic body actions,** organized into **axial movements,** which do not cause the body to travel through the environment in a significant way, and **locomotor movements,** which are aimed at moving a body to a new location in the environment. These basic actions (shown in box 4.4 and figure 4.5) can occur on any body part and can often occur in temporal clusters that are important to resolve with one another.

One possibly surprising omission from this list is "facial expression"; and this reflects the descriptive emphasis of the BESST System. While we might typically or more consciously notice the absence or presence of a smile, what is often more important to consider is the postures and gestures of the rest of the body that accommodate the smile. We need to consider questions like "Was the action of the mouth (expand) accommodated by a slackening (expand) of the face overall and a stiffening (condense) of the spine?" as the answers are usually key to resolving why the change in bodily container could be classified as a "smile" in a particular context. Moreover, this list gives us a new language with which to analyze a given smile. Was the smile a gesture, condensing in a temporal moment to communicate something to the environment, or a posture, part of a longer phrase that was about the internal support of the mover? By not jumping to a limited list of possible expressions or other coherent, stereotyped actions, like "wash," "dry," or "paint," which are limited in their application to a particular task—or even the names of movement from codified systems of expression, like "pirouette," "triplet turn," or "rock step," which are refined combinations of these basic actions—we can resolve full bodily expression in more detail across many more contexts.

Embodied Exercises

- **Basic body action scavenger hunt:** Look for each of the distinct body actions over the course of the artist Robyn's *Call Your Girlfriend* music video (https://www.youtube.com/watch?v=Nv644ipg2Ss). (Most of them appear.) Where do you find them? How much of the performer's body do you need to see in order to identify them? Which actions do not appear for you?

- **Choreography with basic body actions:** Create a floor pattern map using the symbols provided in this section to create a movement design; enact this design in your own space using your informal notation. (See the example provided in figure 4.6.) How does moving this phrase bring new nuance and more complexity to the task of notating it? What movements did you need to add to connect the ideas you sketched out? Did your floor pattern accommodate the actions you drew?

Box 4.4
Basic Body Actions

- **Axial movements:** These are changes in the container of the body that do not involve translation in space.
  - Change of support (act)[9]
  - Posture[10]
  - Gesture
  - Condense
  - Expand
  - Rotate[11]
  - Vocalize[12]
  - Focus
  - Touch[13]
  - Hold
- **Locomotor movements:** These are complex, often cyclic repetitions of a series of axial movements that support the translation of the body in space. The list here refers to bipeds with two additional limbs (arms) available for weight support. For bodies with more limbs, additional modes of locomotion (e.g., pronking for quadrupeds) are available.
  - Change of support in series (act in series, or travel)[14]
  - Roll
  - Slide
  - Crawl
  - Walk
  - Run
  - Jump
    - 1 to 1 (one foot to the same foot, e.g., a hop, where the change in weight typically, but not always, occurs on the supporting surface of the foot)
    - 1 to 1 (one foot to the other, e.g., a leap or *jeté* in ballet)
    - 1 to 2 (e.g., an *assemblé* in ballet or in the long jump event in track and field)
    - 2 to 1 (e.g., a *sissonne* in ballet)
    - 2 to 2 (both feet to both feet, e.g., as often used when shooting a basketball)

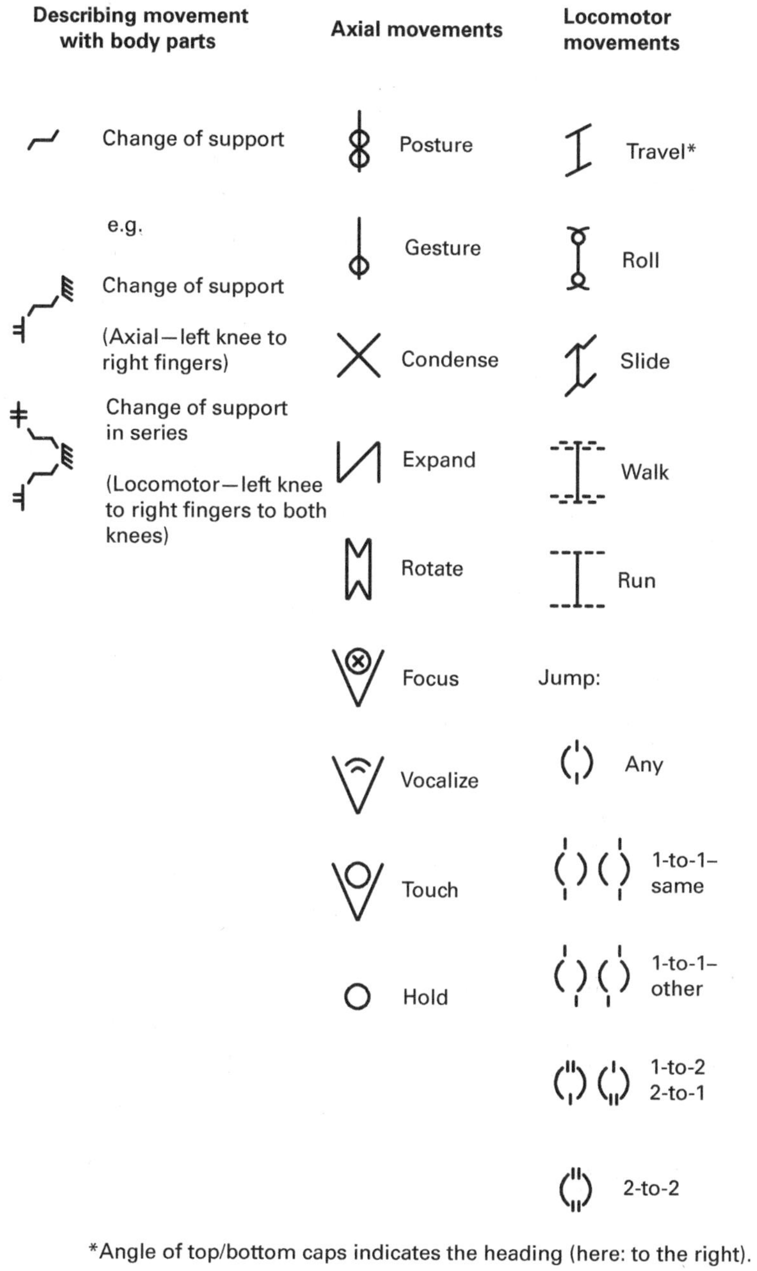

Describing movement with body parts	Axial movements	Locomotor movements
Change of support	Posture	Travel*
e.g. Change of support (Axial—left knee to right fingers)	Gesture	Roll
	Condense	Slide
Change of support in series (Locomotor—left knee to right fingers to both knees)	Expand	Walk
	Rotate	Run
	Focus	Jump:
	Vocalize	Any
	Touch	1-to-1– same
	Hold	1-to-1– other
		1-to-2 2-to-1
		2-to-2

*Angle of top/bottom caps indicates the heading (here: to the right).

Figure 4.5
Symbology for basic body actions.

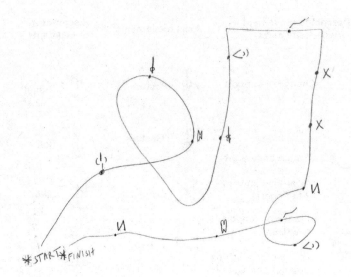

Figure 4.6
Sample floor pattern map with basic body actions along the way. This shorthand
allows movement description and arrangement in space and time.

4.4   Body Fundamentals
An outline of somatic support for movement

Body (or **Bartenieff**) **Fundamentals (BF)** include both **BF Principles** and the
Basic Six (introduced in the embodied exercises at the end of this section).
BF Principles are large ideas of connectivity and support for movement
intention and experience. These principles, listed in box 4.5, are considered an unordered palette of concepts that help bring individual movers
into a better relationship with their own movement patterns. They are
interrelated and have a part/whole relationship to both the mover and
the mover's intention, as well as to the efficiency and connectedness of the
mover's body.

Embodied Exercises

The exercises for this section are a canonical set of patterns established in
BF to help demonstrate the principles outlined in this section. The Basic Six
is a set of movement ideas, sequences, and patterns developed to get to the

Box 4.5
BF Principles

Spatial and Temporal Relationships

- **Dynamic alignment:** Our body design is a closed system, and a change to one part will change the whole. This idea of skeletal connections supports access to kinetic chains of movement, framing a key idea that every and any motion in the body affects all of the body.

- **Axis of length:** Our human form is composed primarily of length. We are typically longer (taller) than we are wide or deep. This principle gives us access to that sense of verticality and aligns with the pattern of our bilateral symmetry. Finding thoroughness and connectivity—which is experienced through the center (or midline) of the body (head-tail-heels of the feet) and our relationship to gravity, where the head and tail are simultaneously limbs and core—will support our movement in function (e.g., accessing this idea can help improve stability when attempting to isolate the upper body from the lower limbs) and expression (e.g., taking a more upright posture may signal a sense of power and authority).[15]

- **Body-level phrasing:** This deals with the initiation and sequencing of our movement to help clarify intent. "Initiation" describes where the movement begins (a spatial location signified by body part), and "sequencing" describes how the movements are ordered in time.

  - Initiation
    - Core
    - Proximal
    - Midlimb
    - Distal
  - Sequencing[16]
    - Simultaneous (parts moving at the same time)
    - Nonsimultaneous
      - Successive (adjacent parts moving one after the other)
      - Sequential (nonadjacent parts moving one after the other)

Functional Support

- **Breath support:** Breath is fundamental to our existence and can be either unconscious or conscious. In this principle, we activate volitional breath to support our sense of inner volume in connection to the world.[17]

(continued)

Box 4.5 (continued)

- **Core support:** The trunk of the body, which includes the spine and pelvis, is the place where movement can originate, activate, and support our successful negotiation of our relationship to gravity.

- **Rotary support:** Access to complex, three-dimensional movement is supported by the rotary capacity of our joints, which we can cultivate and articulate.

- **Weight support (and shift):** Mobilization, particularly in our relationship to gravity, enhances our efficient locomotion in space.[18]

- **Developmental pattern support:** There is a predicated, patterned progression of limb/core integration in our human motor development. We can access that progression as a support for our adult movement expression. This progression is the underpinning or building block for our ability to progress from infants primarily resting on the floor to being able to eventually stand upright and locomote. Accessing all of these patterns, as we do as adult movers, as well as returning to and isolating earlier patterns, is a practice that can increase our ability to exhibit complex movement in the environment.

Expressive Intent

- **Spatial intent:** Our spatial goals in our movements, explicitly related to the concepts discussed in detail in chapter 5 but foregrounding the bodily, physical experience that creates these more complex ideas.

- **Temporal intent:** Our temporal goals in our movements, explicitly related to the concepts discussed in detail in chapter 6 but foregrounding the bodily, physical experience that creates these more complex ideas.

- **Shape intent:** The change in our form as we move, both our own form and our form in relationship to the environment, explicitly related to the concepts discussed in detail in chapter 7 but foregrounding the bodily, physical experience that creates these more complex ideas.

- **Effort intent:** Our motivation to move as manifest in movement quality, explicitly related to the concepts discussed in detail in chapter 8 but foregrounding the bodily, physical experience that creates these more complex ideas.

Box 4.6

The Basic Six

- **Thigh lift:** Focuses on hip flexion and the iliofemoral relationship of the lower limb to the core. This idea is traditionally demonstrated starting in supine hook position and grounding one foot into the floor while lifting the opposite knee toward the nose and then returning it to the ground. This action supports ease and efficiency in locomotion and level change.

- **Sagittal pelvic shift:** Focuses on shifting the pelvis forward and slightly up in a core-to-limb pattern. It is traditionally demonstrated starting in the supine hook and slightly lifting the pelvis off the floor and shifting it toward the heels. This action supports forward and backward locomotion, level change, and centering of the weight over the base of support.

- **Lateral pelvic shift:** Focuses on a sideways mobilization of the pelvis in a core-to-limb pattern. It is traditionally demonstrated from a supine hook position by slightly lifting the pelvis off the floor and shifting through the trochanters from side to side. This action supports efficient lateral weight shifting.

- **Body half:** Focuses on differentiating the right and left sides of the body by connecting the upper and lower halves on each side. Demonstrated by lying on the floor in a big *X* and through lateral spinal flexion, bringing the knee and elbow toward each other on each side. This action supports finding a clear spinal midline through the body.

- **Knee drop:** Focuses on finding a twist of the lower unit in relationship to the upper unit through rotation. Demonstrated from the supine hook position, the knees drop (passive) or reach (active) to one side while the opposite side of the upper body rotates and extends away from the knees. This action supports finding the connection of upper to lower through a cross-lateral patterning.

- **Arm circle:** Focuses on finding the rotational capacity of the shoulder girdle and arm. Demonstrated from the knee drop position, the arm circles up over the head and down across the pelvis, tracing a circle in both directions with the eyes and head tracking the path of the arm. This action supports access to three-dimensional space from the upper limb to the core.

Explore each of these movement patterns with different elements from the list of BF Principles for different experiences to deepen your understanding of how these ideas enact in practice. For example, how does your experience of thigh lift change when you focus on your breath support instead of on your spatial intent of tracing an arcing pathway of the knee toward the face?

essential level of body connectivity and the mover's conscious awareness of patterns of body-level connections (Bartenieff, 1980/2013). The sequences—listed in box 4.6—can be used for repatterning, diagnosis, full body expression, one-on-one movement work that utilizes touch between client and practitioner (so-called hands-on therapy or practice), among other things. The goal is to facilitate, for the mover, total body integration and connectedness to support efficiency in moving in our uprightness in relation to gravity. For example, Bartenieff used these in her work to reintegrate the upper and lower bodies of both ballet dancers and polio patients. While we present the form in which Bartenieff described these sequences, they are meant to be explored in many configurations, contexts, and relationships to gravity.

4.5   Application to Machines: Generating Artificial Gait
Skittering, sauntering, and staggering

The Basic Six and supporting principles form an embodied basis for thinking about complex movement, including gait. Here, we have focused on gait through three sequences of the Basic Six, sagittal and lateral pelvic shift and thigh lift, as Bartenieff did in her work with polio patients and ballet dancers. This puts the emphasis on the role of the spine in walking, which is downplayed in biomechanical analyses that focus on the large deformation (and corresponding muscle activity) that occurs in the lower limbs during bipedal gait (Cenciarini & Dollar, 2011). The motion of the spine has also been studied in walking, but the motion is so subtle that motion-capture markers have to be surgically implanted (Crosbie et al., 1997). The Basic Six and embodied investigation allows us to sidestep that measurement challenge, instead utilizing a somatic strategy in our design of an artificial walker.

As shown in figure 4.7, working with Umer Huzaifa, we associated these ideas with a planar walker, where a simplified idea of sagittal pelvic shift and thigh lift are mapped to movement parameters (*PS* and *TL*), embedded as variable constraints, along with other fixed constraints required for a walking gait, in a feasibility formulation of an optimal control problem, solved numerically in MATLAB (Huzaifa et al., 2020). This produces hundreds of distinct gaits, six of which have been validated in online user studies in Amazon Mechanical Turk, with users rating the gaits on a scale from 1 to 7 after being trained on these distinct gait terms via selected

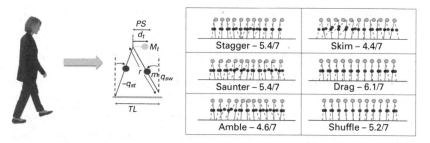

Figure 4.7
Results from robotic gait development. In our research, we have used ideas from embodied movement studies as somatic strategies to generate varied artificial motion and biomimetic platform design. At left, the mapping between the human form and a simplified model (Huzaifa et al., 2016). At right, unpublished results showing ratings by human subjects from Amazon Mechanical Turk experiments (UIUC IRB #17697). Similar results have been published for a simpler bipedal model (Huzaifa et al., 2020). Figure by Umer Huzaifa, used with permission.

definitions. Although about sixty synonyms for "walk" were found in the English language, these six have been validated alongside quantitative planar gait models, using the definitions provided in Huzaifa et al. (2020):

- **Drag:** to trail, to hang with its weight, while moving or being moved; to move with friction on the ground or surface.
  - *I have to drag myself out of bed each day.*
- **Lope:** to run or move with a long, bounding stride.
  - *She put the horse into a lope and headed for the shed.*
- **Saunter:** walk leisurely and with no apparent aim.
  - *In June, some flights were delayed at the airport when about 100 turtles, seeking a place to lay their eggs, sauntered across a runway.*
- **Shuffle:** walk by dragging one's feet along or without lifting them fully from the ground.
  - *I stepped into my skis and shuffled to the edge of the steep slope.*
- **Skim:** to move, glide, fly, or float lightly and rapidly over or along (the ground, etc.).
  - *The swallows skimmed along the surface of the water.*

- **Stagger:** to sway involuntarily from side to side when trying to stand or walk erect.

  ◦ *A young woman staggered toward the landlady and fell down in a swoon.*

This research with BF demonstrates how external measurements like displacements do not align with the perceptual landscape contained within embodied frameworks. Moreover, the production of new movement patterns—along with hardware designs (Huzaifa et al., 2016, 2019)—represent an innovation that may improve bipedal walking behavior in future humanoid platforms. Follow-on studies have further explored the efficacy of various label sets across a variety of contexts (Lambert et al., 2019); generally, we find that emotive labels resonate more with lay subjects but vary more across contexts. These variable walking styles, when visibly distinct for human counterparts, may allow an ongoing advertisement of aspects of the robot's internal state to people in parallel with other necessary tasks, creating more expressive robotic systems for more harmonious integration of artificially embodied machines in human-facing environments.

Embodied Exercises

- **Breaking down gait with BF:** Try to see if you can identify a (typically simultaneous) sagittal and lateral pelvic shift, as well as thigh lift, in your own gait. Play with exaggerating and diminishing the magnitude of these components. What kinds of gaits emerge with an exaggerated sagittal (or lateral) shift? Are these useful in certain contexts? How would you modify your shifts if you were walking on ice? If you move differently from the kind of biped we examine here—for example, with a walker or wheelchair—do these concepts map to your experience of moving, or not?

- **Trying an artificial bipedal gait:** Watch a video of artificial gaits from our research (https://www.youtube.com/watch?v=pDTUeP5S2FU) and try them with your own body. What new movements do you have to invent (using your more complex form) to create similar temporal patterns? How well do the verbs assigned in this research fit with your experience?

4.6 Exploring the Themes: Body through the Lens of Stability/Mobility
Moving to hold still

One way to begin to experience the moving body is to understand the interconnected relationship of stability and mobility. While they may seem to be opposite states of movement, they are simultaneous and ongoing patterns present in all movement. In this context, "stability" is different from "immobility," which implies a holding, stuck and unmovable. Stability supports mobility and vice versa.

Take, for example, a right-handed tennis player executing a backhand shot. The arms are mobilizing but can only be mobilized with the force of the swing by stabilizing the lower body. Likewise, in preparation for the backhand swing, the lower body mobilizes toward the ball while the upper body stabilizes in response to the reaction of the racket against the ball. This relationship—also called the Stability/Mobility theme—can also be seen in more pedestrian actions. To achieve a change of support from one foot to the other as in walking, the right side of the pelvis, right leg, and right foot mobilize, while the left side of the pelvis, left leg, and left foot stabilize to support the push/reach forward into space. This relationship between mobility and stability can also be clearly seen in muscle action: flexion of the elbow requires the contraction or stabilization of the biceps, while the triceps lengthens and mobilizes. The ongoing balance of the two in relationship promotes the ability to move the body.

The idea that you have to move to stabilize and you need underlying stability to move with successful control are principles in feedback control of dynamic systems as well. Feedback control does not offer the topology of duality in highlighting the simultaneity of these ideas, but they are important for human movers, who may overdo the movement in one direction or the other. As Nettl-Fiol and Vanier (2011) write in their book applying Alexander Technique to dancer training, patterns of holding can be counterproductive to increasing the virtuosity of movements like *grand battement*, where dancers need to release the working leg while supporting the standing leg in stability. Clearly, this duality appears in other traditions of movement studies and is useful for embodied practitioners to consider when refining their own movement choices: to perform a complex turn, we need a stable shoulder girdle; to balance on one foot in stillness, we need supple activation of the sole of the foot and the pelvic floor. That is, the body is a dynamic site of perceptual exchange where some movements are called "stable" and some "mobile" (and in another context, these labels may be switched), but both contribute to the eventual embodied existence of movement.

Embodied Exercises

- **Exploring mobility and stability in locomotion:** Locomotion is a highly mobile task, moving a body from one place to another, but it cannot be accomplished (especially for bipeds, which are inherently unstable forms) without stabilization.

  - Try walking at an easy pace, and then accelerating into a run.
  - Now freeze and see if you can come to stillness.
  - Balance on one leg, and then the other.
  - What happens when your body is forced in and out of motion? What do you notice about your musculature? Can you notice what parts are mobilizing and what parts are stabilizing in the sequence of movements given here? Can you notice the tiny movements that your body uses to hold "still"?

Chapter Summary

This chapter has described several concepts of embodiment from the Laban/ Bartenieff tradition that have been useful in our work, codified as the Body component of the BESST System. We have shared anatomical relationships as perceived in the process of moving that help clarify movement patterns. We have given names to the experiences of *being* and *doing* through the sensations of weight and flow. We have suggested a taxonomy for naming movements that does not belie the purpose or intent of those movements and named principles of bodily support that enable such complex movements. In addition, we have shared how this mode of investigation has affected our own quantitative exploration into variable styles of walking gaits. Finally, we reviewed a common duality named in LBMS (stability/ mobility) through the lens of Body. Thus, our physical container and our perception of it alone are not enough to establish our complex ability to move. *Both* body organization (physical/function) and movement intention (mental/expression) combine to create successful movement in an ongoing, fluctuating relationship of stability and mobility of parts that enact a whole.

5   Where Is the Movement? Spatial Fiducials for Movement (Space)

Just as chapter 4 provided answers to the question of *what* is moving, this chapter examines the *where* of the phenomenon of movement. It can seem an obvious or rather simplistic topic to cover in fields like engineering, animation, and design, where vector spaces and coordinate systems are well established and frequently used, especially with computer-aided design tools. In other words, quantitatively locating objects in space is a well-studied problem that the field of movement studies does not help solve. What this system aims to highlight is *how bodies move through space* and *how that movement creates meaning in a given environment.* We denote then, the Space component as distinct from (but related to) the concept of space as typically used in engineering and design. Engineered systems need to resolve relative approaches and attitudes toward space (the topic of this chapter) as well as measure quantities of distance (a topic well-covered in traditional textbooks in robotics).

For example, objects located four feet from one another can be absolutely measured well by engineering terms, but whether two bodies inhabit the same level of their "kinesphere" (a concept introduced in this section) is not about absolute measurement, but rather about the relative attitude toward each body's approach to the space that the objects inhabit. For a relatively large body, like that of a professional basketball player, to *reach up*, for instance, the distal ends of its fingertips may be eight feet off the ground; for a relatively small body, like that of a child, to take a similar action, the distal ends of the fingertips may be only four feet off the ground. Both bodies can reach and stretch and yearn to move toward the sky as much as they can—creating a similar attitude toward their environment, but with very different physical results. While each body then inhabits a different physical space, the movement reveals a similar approach to space and is consequently read as a similar action.

Today, soldiers operating simple mobile manipulators like the PackBot, a robot on two rugged, crawler-style augmented wheels with a four-degree-of-freedom manipulator arm and a two-degree-of-freedom gripper with camera attached, struggle to navigate the robot successfully through situations that the hardware itself can handle. For example, having the base translate at the same time as the arm is not accommodated by current control interfaces but could be accomplished by the set of onboard actuators and sensors. This suggests a breakdown in command architecture and a lack of shared spatial awareness between the human operator and the artificial device.

This forms a fundamental question that sits at the nexus of robotics and movement studies: how can we map one body's understanding of space to another, wildly distinct body? Such translation, between human and robot kinespheres, can be seen as the central task for human robot teleoperators. While it is well understood how to move robot manipulators into different spatial positions (converting between joint space and task space through geometric relationships), it is not yet clear how to efficiently communicate control of a robot body from a human one. Some ideas, like exoskeletons for mapping motion through a wearable armature that aims to capture human movement directly, have shown initial progress, but there persists an inability to improvise and fully utilize the physical capability of the remote robotic device.

5.1   Approach to the Kinesphere

Defining an abstraction to describe the space we can move in

As part of the larger environment of the general space we move in, we each have an amount of personal space around our own bodies—what we call our **kinesphere.** This space is defined by "the sphere around the body whose periphery can be reached by easily extended limbs without stepping away from that place which is the point of support when standing on one foot" (Laban, 1966, p.10). In other words, the kinesphere is delineated by how far we can reach around ourselves. Beyond this physical sense of space, the term "kinesphere" is also discussed as a broader metaphor for the idea that movement from a given body interacts with space in a way that is observable. We also recognize an **innersphere,** or the idea that the body has space inside it that, while not visible, can be experienced and accessed in supporting the body expression of space, as well as a **coronasphere**,[1] or the idea that our breath accesses a farther-flung volume of space. That is, we perceive space around familiar moving bodies as capable of being filled with their presence. These terms are listed and further described in box 5.1.

When we move in space, changing our position in the environment, our kinesphere always comes with us. In drawing attention to this volume, the BESST System names a space where a human body can be expressive, where a given human can do functional work. However, purely considered as a volume, the area reachable by a given body is uniquely shaped by that body—and may change over the course of that body's lifetime, or even over the course of a day. We can refine and define our understanding of the volume of the kinesphere by dividing it into parts such as **zones, levels, reach space,** and **pathways** to finesse a greater access and range of motion as a mover. Thus, the lists in boxes 5.2 and 5.3 (and the symbols in figure 5.1) are several ways that we can identify, observe, and interpret this approach to the kinesphere, all of which are understood relative to the body of the mover.

Zones and levels are useful ways to delineate the volume of the kinesphere, with the key idea being that the area in which a movement occurs

---

Box 5.1
Defining Spatial Spheres of Human Movement

- **Kinesphere (personal space):** Recognizes that our body has contents outside the container of the skin and within our physical reach. Access to experiencing this sphere is often accomplished through the use of varied movement exercises, such as the "movement scales" introduced later in this chapter, which help broaden our repetitive movement patterns in order to find areas of this sphere we may not typically, but still can inhabit.[2]
- **Innersphere (inner space):** Recognizes that our body has contents inside the container of the skin, which include our own inner volume of space. Access to experiencing this sphere is often accomplished through focusing on the internal sensation of breath.
- **Coronasphere (breath space):** Recognizes that our body has contents outside the container of the skin and beyond our physical reach. Access to experiencing this sphere is often accomplished through imagining the external expression of breath.

Box 5.2
Zones, Levels, and Reach Spaces

- **Zones and levels:** Denote large areas of space relative to the body.[3]
  - Levels of verticality:
    - High level (above the head)
    - Middle level (around the waist)
    - Low level (around the feet)
  - Sidespace (both sides of the body):
    - Right zone (whole right side)
    - Left zone (whole left side)
  - Sagittal dissections:
    - Front zone or frontspace (whole frontal area)
    - Back zone or backspace (whole back area)
- **Reach spaces:** Where movement occurs relative to the core of the body.
  - Near-reach (close to the body)
  - Mid-reach (between the body and the edge of the kinesphere)[4]
  - Far-reach (at the edge of the kinesphere)

Box 5.3
Pathways

- **Central pathways:** Follow a trajectory toward and away from the center of the body in a somewhat linear fashion.
- **Peripheral pathways:** Trace along the edge of the kinesphere, maintaining a constant distance between the center and the articulated edge in any reach space.
- **Transverse pathways:** Modulate between the center and the edge of the kinesphere, negotiating the in-between space.

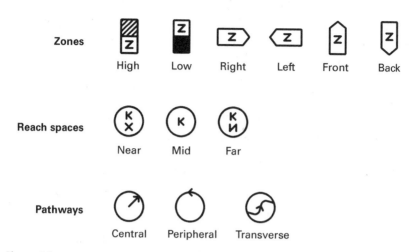

Figure 5.1
Symbols for kinesphere: zones, reach spaces, and pathways.

supports and/or delineates both its form and function. For example, a short person trying to get the attention of a friend in a crowd of people might use the high level of his kinesphere to wave his arms around so that the movement can be seen over the heads of the crowd. This action is needed to serve both functional (to avoid occlusion by the heads of other people) and expressive (to communicate with the friend) goals. A very tall person might only need to use the middle level to wave her arms around and attract a friend's attention because her height already allows arm gestures to be seen across a crowd, but this person still may use the high level and far-reach space of her kinesphere to serve the expressive goal of communicating urgency. The high level and far-reach space are still useful in this instance, even though they are not functionally necessary. Thus, the relationships between reach spaces and zones and levels are personal to each mover, and various tasks require different usages of them.

Pathways can be understood as the route that a movement takes through the kinesphere. As the body moves (both in terms of gestural and postural changes), choices about the trajectory may be apparent, distinct from locomotion, focus, or posture. The Latin word *trajectoria* means "throw across," so it is possible to imagine pathways as trajectories "thrown across" the mover's kinesphere in different ways. The process of that pathway unfolds over time and can be seen as relating to away, toward, or through the center; along the edge; or the space in between.

As an approach to the kinesphere, pathways serve as both expressive and functional choices. When giving someone directions (using hand and arm gestures to describe the route), the use of central and peripheral pathways might be better at describing getting from here to there (central) along a road (peripheral). Transverse pathways in this example might be confusing and unclear for delineating a route to follow. A basketball player (while handling the ball) might utilize central pathways to pass the ball to another player, while using transverse pathways to dodge, block, and feint (while holding the ball) in order not to lose the ball, and ultimately might engage a peripheral pathway to make that long, three-point shot.

Embodied Exercises

- **Initial exploration of the kinesphere:** In this exercise, you will explore the options of movement location around your body without changing your location in space.

  ○ Begin by standing in a neutral stance. Explore the space above your head. Do you automatically use your arms? Can you move your legs into this space? What associations does movement in this space engender? Try the same type of exploration for the middle and low levels. What feels familiar? What sort of actions do you think you do in these different spaces? Which level do you experience more fully with arms, legs, or core? How are they the same and different?

  ○ Do the same exploration for frontspace, backspace, and sidespace. What sort of tasks seem to be supported in these different zones? Do you experience a preference or familiarity with one zone over another?

  ○ Continue to explore the space around your body that defines your kinesphere. Make any notes about what you discover about both functionality and meaning. What space do you feel you have the greatest access to? Why do you think that is the case?

- **Using images to find meaning in space:** This exercise engages different attitudes toward space via distinct imagistic descriptions that invite exploration.

○ Lie on the floor and feel everything drain out of you and into the earth. Grow roots down into the earth—how can you move and feel this connection? What if you were moving on the bottom of the ocean, in a muddy swamp, or on a rocky hillside?

○ Try moving toward your feet (down or forward), head (up or backward), and sides (right and left) on your belly and on your back. How does your experience change as you use the front zone of your body? The back zone? The sidespace? What changes occur in each of these iterations?

○ Gradually work your way to sitting, and then standing. What happens as you change your level in the space?

○ Notice how we manipulate the same ideas in Space through familiar images in the opening of this exercise versus more literal, explicit directions or tasks in the previous exercises.

• **Transforming body actions:** In this exercise, you will explore transforming the same basic body action through different concepts in Space.

○ Zones and levels:

▪ Manipulate moving an object around on your desk in the front zone. Now try manipulating the same object in the right or left zone. How does your experience of the object change? What could this shift be useful for facilitating?

○ Reach spaces:

▪ Enact back-and-forth lateral action of the open, outward-facing palm in three different reach spaces.

• Wipe your nose (near); wave hello (mid); ward off unwanted attention (far).

○ Pathways:

▪ Pick up a pencil using each kind of pathway.

• Central (efficiency)

• Peripheral (inefficiency or distraction)

• Transverse (complexity, even emotional)

Notice which pathways seem to make sense to you for the task at hand, and how changing the pathway changes the meaning/experience of the movement.

5.2 Spatial Pulls and Platonic Solids
Spatial ideas that help organize and analyze movement

Punctuating the kinesphere are notable **spatial pulls** (listed in figure 5.2). While engineers tend to model space as an infinite continuum of possible directions, represented in Cartesian coordinates by three orthogonal real number lines, humans cannot perceive the difference between a vector pointing to (0,0,1) and (0,0,0.999); moreover, in the environments that humans typically inhabit, this difference is not often *meaningful*. Thus, the spatial pulls discussed in the Space category of the BESST System are a select and small subset of possible directions—perceptual fiducials that are useful for human observation. Because we have a sense for how a person might, can, or will move in the space around the self, the choices that a particular mover actually makes in the space are expressive. This is often described as a dynamic tension between the mover and the environment. It is most acutely felt as an ongoing interaction: rather than a list of absolute or even relative spatial locations, a spatial pull describes this relationship as indicated through the moving body. Thus, the process of change is what becomes meaningful rather than a set of coordinates or "places."

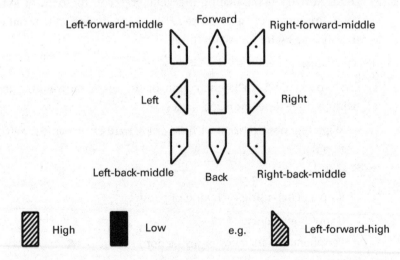

Figure 5.2
Symbols for spatial pulls. The nine shapes for the middle level (shown at the top and middle) can be shaded to indicate the high and low planes (shown at the bottom).

If one body inhabits the upper level of its adjacent area, it does not have to be *as far* up as another body to still be *up*. Likewise, a particular style of reach upward can actually intone a downward spatial attention, creating a different action entirely. Thus, we note that these twenty-six spatial pulls can be used to divide the levels and zones previously introduced, providing finer-grained, more detailed abstractions for seeing similarities in distinct bodies navigating space. In doing so, we name *dimensions, diagonals, planes,* and, later, Platonic solids (*tetrahedron, octahedron, cube, icosahedron,* and *dodecahedron*) as models of the kinesphere. These elements (see the symbol sets in figures 5.3 and 5.4) combine actions along the dimensions that are roughly aligned with the principal axes of inertia of the human form and are recognized as relative to the mover (up is always toward the head and down is always toward the feet). Boxes 5.4, 5.5, and 5.6 list the architecture of three of these models, the octahedron, cube, and icosahedron, in terms of dimensions, diagonals, and planes, respectively.

Vertical   Sagittal   Horizontal

Figure 5.3
Symbols for indicating planes. Three planes bisect the mover—vertical, sagittal, and horizontal—creating opportunities for movements of different anatomical emphasis.

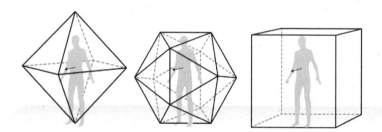

Figure 5.4
Models for a mover's kinesphere. Here, moving left to right, the kinesphere is imagined as an octahedron (with inscribed dimensions), icosahedron (with inscribed planes), and cube (with inscribed diagonals). The geometric forms, as well as the figures inside, are tilted on an angle (as indicated by the black arrows) to better show their spatial relationship.

Box 5.4
Spatial Pulls in the Octahedron

- Pure spatial pulls in the octahedron (formed by three orthogonal dimensions)
  - Vertical dimension (major axis of length)[5]
    - Middle[6]
    - High
    - Low
  - Sagittal dimension (minor axis orthogonal to length)
    - Middle
    - Forward (sometimes called "front," defined by a sensory-rich face that implies a preferred heading for travel)
    - Back
  - Horizontal dimension (minor axis orthogonal to length)
    - Middle
    - Right
    - Left

Box 5.5
Spatial Pulls in the Cube

- Combinations of spatial pulls in the cube (three evenly felt spatial pulls, which form diagonals)
  - Right-forward-high
  - Left-back-low
  - Left-forward-high
  - Right-back-low
  - Left-back-high
  - Right-forward-low
  - Right-back-high
  - Left-forward-low

Box 5.6
Spatial Pulls in the Icosahedron

- Combinations of spatial pulls in the icosahedron (two unevenly felt spatial pulls)
  - Vertical plane (vertical dimension is primary and horizontal dimension is secondary)
    - Right-high
    - Left-high
    - Left-low
    - Right-low
  - Sagittal plane (sagittal dimension is primary and vertical dimension is secondary)
    - Forward-high
    - Forward-low
    - Back-low
    - Back-high
  - Horizontal plane (horizontal dimension is primary and sagittal dimension is secondary)
    - Right-forward-middle
    - Left-forward-middle[7]
    - Left-back-middle
    - Right-back-middle

The infrastructure of the octahedron is the dimensional *cross of axes*, which defines the vertical (height), sagittal (depth), and horizontal (width) dimensions. These orthogonal dimensions can be thought of as rather straightforward body frame coordinates that emanate from the center of mass of the mover. However, it is worth discussing the primacy of the vertical dimension (which is sometimes named the "longitudinal axis"). When standing upright, this dimension aligns with the downward pull of gravity. It is also the axis about which humans find it easiest to spin (it corresponds to the smallest moment of inertia in most adult humans). And, as referenced in the Body component, this verticality is one of the first things we look to in order to understand a foreign moving body (whether human or not). Puppeteers turn strips of foam into expressive characters by, among other things, creating a consistent relationship to gravity for the puppet. While all bodies can change that relationship, the vertical dimension remains a primary factor in our experience of movement in space (both our movement and the movement of other bodies).

The infrastructure of the cube consists of four diagonals that define three equal-dimensional pulls expressed simultaneously as one diagonal direction, with its equal and opposite pull at the other end. These three equally felt pulls create a sense of mobility and off-balance reactions in the body, revealing a key relationship between the Body and Space components. Often, it is observed that reacting to these spatial pulls creates (or requires) a complex spiral through the body. For example, the scapula of the arm indicating or responding to the pull will literally rotate down and around the shoulder joint in order to afford greater access to space by the arm. Moreover, shifting weight through the lower body can continue this spiraling action through the pelvis to the pads of the feet. These bodily changes can create mesmerizing patterns in the human form, particularly with performing artists who are trained to emphasize and reveal these changes to observers. For example, the diagonal directions are often leveraged in proscenium-based performing arts like ballet, where performers will indicate the corners of the room through different bodily positions such as *epaulé derrière*, the body position where the right arm crosses in front of the body and rotates toward left-forward-high, while the left arm crisscrossed behind to right-back-low and the legs follow to reveal a similar facing.

The infrastructure of the icosahedron is the three planes: vertical, sagittal, and horizontal (each having two unequal pulls emphasizing the correspondingly named dimension). We often describe the vertical as the **door plane,** stable (as the human form is when stretched out, upright, in an *X*), with more up and down than side to side; the sagittal as the **wheel plane,** mobile (as the human form is when attempting a forward step or somersault), with more forward and back than up and down; and the horizontal as the **table plane,** a place most associated with manipulation (or expression) by the hands and arms, with more side to side than forward and back. These directional pulls are thought to be more natural, occurring often in everyday life, compared to the complexity of the diagonals in the cube or the austere purity of the single-dimensional pulls in the octahedron.

As we will explore further in the next section, moving among these spatial pulls creates immediate implications of geometric change, implying something inherent in the expressive nature and functional associations of each movement. Movements in the cube can be dramatic changes reversing three polarities (e.g., between right-forward-high and left-back-low) or off-balance actions that cause complex bodily accommodations (e.g., left-back-low to left-forward-low). Movements in the octahedron, on the other hand, require rigid severity in the purity of each pull (e.g., moving from back-middle to forward-middle without any hint of change in high to low or left to right). This pull, first, is an abstract ideal that we cannot achieve, and second, entails an unnatural strictness that can be tiring. The icosahedron is thought to invite much more natural movements. That is, gliding between the two unequal pulls in this form may be a "Goldilocks" solution to physical movement commonly inhabited in the kinesphere: changing neither too abruptly nor too gradually, but just right.

Embodied Exercises

- 📹 **Finding your own personal, affective associations with space:** Use the prompts here as free movement responses. When responding, think about the following question: Which areas of your kinesphere did you visit more frequently? Jot your responses down on paper and compare across prompts. It may be instructive to record yourself in order to observe separately from moving (or, feel free to observe simultaneously with your movement response).
  - Bright, bubbly
  - Deep, dark
  - Peaceful, placid
  - Exhaustion
  - Jumping for joy
  - Apathetic indifference
  - Catching slippery soap bubbles

In enlivening your kinesphere with your own interpretation of these different ener-
gies, what patterns emerged? How does the form of your body, including your own
range of motion and prior movement experience, inform these patterns?

- 📹 **Indicating versus revealing space:** In this exercise, we will zoom in on
  one spatial pull—right-forward-high, a diagonal—but invite the reader
  to try it with other directional pulls as well. It will be instructive to record
  yourself here too.

  ◦ Keeping a neutral body posture, employing a forward gaze and even
    distribution of weight between two feet, reach your right arm straight
    up, then out to the right, and then forward, aiming for equal mea-
    sures of each change.

    ▪ You are now *indicating* a point in space that is relatively right, for-
      ward, and high of your center of mass. This, however, is not the
      same as *revealing a spatial pull*, where your entire body responds to
      a pull from far outside your own body. In fact, if you half-heartedly
      stood from a chair to complete the exercise, your body may be
      more likely to exhibit an attraction to that chair than to the
      right-forward-high diagonal, as expressed through a downward,
      slouched relaxation in your lower half.

  ◦ From a neutral body posture, employing a forward gaze and even dis-
    tribution of weight between your feet, imagine that a valuable, price-
    less object of great personal sentiment is located just out of reach
    along the same right-forward-high diagonal that you indicated in the
    previous step. Now, reach for that object. How is this end state, with
    your arm reaching to a similar location, different than in the previous
    step? What happened to your gaze? What happened to your posture?

    ▪ If you are reaching with every ounce of your being toward this
      imaginary object, you are likely recruiting your gaze, a shift of
      weight, the deformability of your torso, muscular tension, and
      other features of your movable body to *reveal* the right-forward-
      high diagonal. Whereas in the first step, your shoulder socket likely
      remained neutral, with the scapula at rest on your back, now your
      shoulder is probably experiencing a rotation, with the scapula slid-
      ing down, maybe even beginning to wrap under your armpit, in
      order to facilitate a more engaged reaching action of your body.
      This reach does not end—because the object is out of reach—but

an observer will now be able to *perceive* where you are reaching with greater clarity.

∘ Try to find the same diagonal (a spatial pull in the cube) from the octahedron (reaching first to high, then right, then forward) and the icosahedron (reaching first to right-high and then toward forward-high, without losing the right component of your intention). Which form is easier for you to start at in order to find the diagonal pulls of the cube? Notice how moving between these forms helps—even forces—you to become clearer in your movements (and your perception of them).

We encourage you to revisit these exercises in chapters 7 and 8 to further analyze the changes in your body used to reveal this direction, understanding them as part of Shape and Effort.

5.3 Movement Scales
Identifying and practicing patterns in space

In much the same way that music is practiced and learned through the study of scales and arrangements of notes in a progression, the BESST System names **movement scales** that map the space of the human kinesphere. These scales are an arrangement of spatial pulls in a progression that, similar to musical scales, are practice sequences designed to hone techniques for accessing—moving in—space. They are not an end in themselves, but a means to an end. Established movement forms use a similar approach to learning the technique of that form through a codified movement vocabulary. For example, ballet requires a barre and a set of actions to be practiced over and over (*tendu, degaje, plié, port des bras*, to name just a few). These sets of movements are designed to enhance the expression of the form. Likewise, they are a means to the end of ballet choreography and performance.

The movement scales provide a way to explore pathways, landmarks, proximities, and spatial pulls, as well as enhance the articulation of the mover in the space. Moreover, they can be used to compare styles of motion to one another (just as different styles of songs may use different keys). As such, they are models of how a mover might access space rather than of the space itself. In the same way that a road map provides a means of navigation, the BESST System scales provide a map to the kinesphere. Using the movement scales to practice spatial movement can reveal and clarify action, in terms of both its functionality for the task at hand and its expressive character.

Practicing and utilizing the scales become important for allowing the mover to experience the different qualities of spatial expression, as well as the functionality of patterns in accomplishing goals and actions in the environment. The scales used in the BESST System emerge from a study of the tetrahedron, octahedron, cube, icosahedron, and dodecahedron, with the main emphasis on the octahedron, cube, and icosahedron. The various scales progress in a manner of either **abrupt** change (all pulls changing), **gradual** change (only some pulls changing), or some combination of both, alternating in a sequence, which reveals the character or feeling of the scale itself. The sense and process of change constitute where the meaning is expressed and the functionality of the space is able to be accessed. All the movement scales are expressed as a "loop" or sequence that repeats itself from beginning to end and consequently can be started and finished by beginning the movement in any pull and sequencing through it.

To illuminate the character of the scale itself, let us compare four particular scales: the **dimensional scale** (using all central pathways), one of the **girdle scales** (using all peripheral pathways), the **diagonal scale** (alternating central and peripheral pathways), and the **A scale** (using all transverse pathways). These scales are described in box 5.7. For a full explanation of all the movement scales, including the spatial pulls in each scale listed in box 5.7, refer to appendix B.

By looking at these particular scales, it is clear that moving through space with certain pulls, pathways, and maps for organizing expression reveals different experiences and, ultimately, meanings. For example, the girdle scale outlines the container of the mover's kinesphere, while the A scale explores the fullness of the contents of the mover's kinesphere. The way in which a moving body uses space can signal intent (particularly when unfolding over differing periods of duration), create boundaries, invite connections, and accomplish the ongoing negotiation of the mover to the environment and self to other. Thus, the use of space becomes meaningful, making it important in human-human interaction and, eventually, human-robot interaction (HRI).

We can begin to use the patterns in movement isolated by these scales to recognize that there can be a harmonic relationship between aspects of

---

Box 5.7

Descriptions of Some Movement Scales

- **Dimensional scale (in the octahedron):** When executed with all central pathways, the dimensional scale presents a body poking out and in as each dimensional pull is expressed, followed by a return to middle before poking out again to the next dimensional pull. There is a duple sort of rhythmic pattern created (out/in, out/in, out/in) that builds a repetitive sense of polarity in the movement expressed.

- **Girdle scale (in the icosahedron):** A girdle scale reveals a different sense of expression as it develops through all peripheral pathways in the icosahedron. A girdle scale forms a sense of "going around" or "encircling" as each spatial pull is ordered in sequence next to a nearby neighbor. This circular expression engenders a sense of alertness and clarity of consciousness. There is a sense of being on "the edge" that defines a container of space, which is in stark contrast to the poking through space that is revealed in the dimensional scale.

- **Diagonal scale (in the cube):** The diagonal scale uses alternating central and peripheral pathways. From right-forward-high (moving with the right side leading) to left-back-low, the mover traverses a central pathway through the center of the cube; from left-back-low to left-forward-high, the mover traverses a peripheral pathway along the face and at the edge of the solid. Moving through the center of the cube and then out along the face creates turbulence between stability and mobility.

- **A scale (in the icosahedron):** The A scale uses all twelve icosahedral directions, connecting them with transverse pathways. It loops around the diagonal of right-back-high/left-forward-low. As a result, the most powerful expression of diagonal space is absent. The sequencing of planal organization creates a defensive feeling, which has been likened to the minor scales of music. While the transverse pathways can be expressed in different rhythmic patterns, the overall feeling of the A scale is one in which the full volume of the kinesphere is revealed due to the rather dramatic changes in spatial organization exhibited by its geometry.

---

movement that layer, integrate, and work in concert in order to illuminate seeing, experiencing, and understanding our relationships both with ourselves and the environment. Rudolf Laban (1966, p.195) called this idea "harmony of movement" in *Choreutics,* and Irmgard Bartenieff (1980/2013) refers to this integration as "coping with the environment" in the title of her seminal text. More recently, Studd and Cox (2013/2020, p.165) describe the practice of movement scales as they encourage the mover to "connect the changing form of the body in its dynamic expression in space". Like harmonies in music, where the design of sounds that delight, intrigue, confound, or otherwise engage the human spirit, the idea of harmony in movement is another way of languaging the idea that human movement is a complex, ubiquitous, and textured design space. Just as color theory highlights triplets of colors that designers can use (or not) to create harmonious interior spaces and rooms, movement studies highlight canonical sequences that can be used as tools for practice, observation, and creation.

Embodied Exercises

- **Exploring spatial pulls in sequence:** In this exercise, you will begin to sequence spatial pulls in each of the Platonic solids to explore the different textures and tones of moving in these spatial forms.
  - Begin by choosing any three spatial forms that you would like (see figures 5.2, 5.3, and 5.4) and move among them.
    - What are you expressing? What are you doing?
    - What does this space elicit for you?
    - How are you transitioning from direction to direction?
    - What type of pathways are you using?
  - Choose three spatial pulls from the octahedron (one-dimensional) and move between them.
    - What sort of pathway are you using to go from direction to direction? Central or peripheral?
    - Can you order your space to allow both central and peripheral transitions?
    - What is your experience of the difference between the two?
    - Do you experience this pathway as stable or mobile?
  - Now choose three spatial pulls from the cube (three-dimensional) and move among them.
    - What sort of pathway are you using to go from direction to direction? Central or peripheral?

- Can you order your space to allow both central and peripheral transitions? What is your experience of the difference between the two?
- What do you experience when moving through a complete diagonal where all three pulls change on a central pathway?
- What do you experience when moving between two diagonal directions when two pulls change along the face of the cube versus through the center of the cube?
- What about moving between two diagonal directions when only one pull changes (along an edge of the cube)? What is your experience of that space?
  - Now choose three spatial pulls from the icosahedron (two-dimensional planal pulls).
    - Do you experience new and different pathways in this form? Can you order your space to go from vertical to sagittal to horizontal and find a transverse pathway?
    - Do you experience this space as somehow happening between the edge and the center of your kinesphere?

For the three experiences given here, consider whether it was difficult or easy to experience space in one form compared to another. Was one use of space more comfortable or familiar to you? What places were easier to access? How did you feel during each pull? Did any images or thoughts come to mind? Notice the images, thoughts, and feelings that you experience in the different Platonic solids and how that elicits varying expressions, as well as a sense of functionality.

5.4 Application to Machines: Generating Humanlike Telepresence
Enabling configuration space control through choreographic technologies

Sharing movement ideas, or commands, with others is a common occurrence. For example, a statement like "Pick that red cup *up* and place it to the *right* of the blue one on the *bottom* shelf" uses many spatial references. The BESST System formalizes this idea, creating a symbolic system for aspects of movement related to Body (see chapter 4) and Space (in this chapter). Research exploiting these two elements of the system has developed a platform-invariant movement specification method and related teleoperation scheme that enable functional and expressive pose (or configuration space) control of articulated robots.

First, an extension was established to the standardized robot description system, the Unified Robot Description Format (URDF), creating labels that *overlap*. While a machine needs to have only a single label for each joint and linkage of an articulated body to unambiguously control it, humans refer to body parts redundantly (as is clear from chapter 4). Thus, these overlapping labels, shown in figure 5.5, allow human specification to better align with machine morphology. In this way, the framework uses a choreographic technology to simulate an artificial embodiment that better aligns with how humans describe their own movements. Next, a user moves the robot through a simple scale, establishing body poses for each spatial pull. These poses are stored in a database that is indexed by parameters inspired from Body and Space component symbols from the BESST System. Users can then specify movement through a simplified form of motif, as shown in figure 5.6.

Evaluating the efficacy of this motion specification scheme was accomplished by checking whether the resulting machine behaviors effectively imitated the original human performance of the movement sequence, evaluated by human subjects (Jang Sher et al., 2019). It is impossible for such an imitation to *physically be the same*; instead, the interest lies in *perceptual similarity*. Likewise, associating these symbols with buttons on a game pad, the scheme can be used in real time to generate improvised, complex movements on robotic platforms as shown in figure 5.7 (Zhou et al., 2019). This scheme has outperformed traditional joint-space control methods in

Figure 5.5
Leveraging concepts of space and body to design robotic algorithms. Left: the Body component supports establishing a redundant labeling scheme, similar to that found in section 4.1 of chapter 4. Middle: the Space component lends the concept of movement scales for providing an architecture of the kinesphere. Right: a database establishes relationships between Body and Space. Modified from Jang Sher et al. (2019), used with permission.

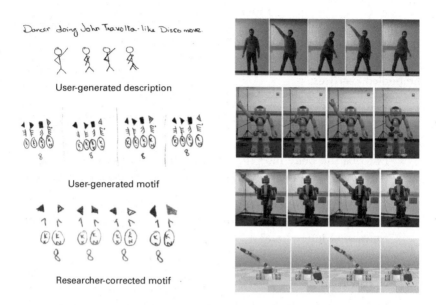

Dancer doing John Travolta-like Disco move.

User-generated description

User-generated motif

Researcher-corrected motif

Figure 5.6
Movement specification through movement notation, which allows execution across distinct robotic platforms. Left: movement notation provided by users and corrected for accuracy by researchers. Right: "the same" sequence on four different bodies, from top to bottom: a human subject, the Softbank NAO robot, the Rethink Robotics Baxter robot, the KUKA youBot. Modified from Jang Sher (2017) and Jang Sher et al. (2019), used with permission.

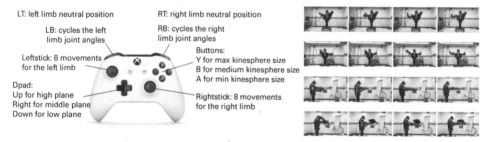

LT: left limb neutral position
LB: cycles the left limb joint angles
Leftstick: 8 movements for the left limb
Dpad:
Up for high plane
Right for middle plane
Down for low plane

RT: right limb neutral position
RB: cycles the right limb joint angles
Buttons:
Y for max kinesphere size
B for medium kinesphere size
A for min kinesphere size
Rightstick: 8 movements for the right limb

Figure 5.7
Leveraging choreographic technology for a joint-space control teleoperation scheme. Left: mapping between Space and Body component elements and buttons on a gamepad, used to control the Baxter robot. Right: snapshots of a human subject operating the robot after a half-hour training session across four different tasks, shown in each row. Modified from Zhou et al. (2019), used with permission.

a variety of tasks that required users to improvise in order to accomplish unknown or unanticipated goals (Bushman et al., 2020).

Embodied Exercises

- **Distinct bodies in shared space:** Consider the three robotic platforms in figure 5.6.
  - Describe the body morphology of each in a paragraph, noting not only how the robots differ from each other, but how each differs from the human body.
  - Which machine has the greatest access to the following levels and zones of its kinesphere?
    - High level
    - Low zone

    Note how the mobile manipulator robot (youBot) has a full range of motion in its high level, whereas the "head" of the small humanoid (NAO) offers interference in this level. On the other hand, only the small humanoid (NAO) can lower itself by bending its "knee" joints, affording some access to the low level.

    - Back zone
    - Front zone

    Note how the large robot (Baxter) can rotate each of its arms 180 degrees, affording full access to both the front and back zones, unlike the small humanoid (NAO), which has limited back zone access (as do humans).

  - How would you express the following spatial pulls on each device?
    - High
    - Right-forward-high
    - Back-low
    - Left-forward-middle
- **Movement specification on distinct bodies:** Use the concepts introduced so far in this chapter to create and describe three movement behaviors. Write your descriptions and then enact them in your own body. See appendix E for a version of this exercise for a group.

5.5   Exploring the Themes: Space through the Lens of Function/Expression

Highlighting the utility of expression

Changes in space affect both the expressive meaning and functional purpose of a movement. In the BESST System, this principle is understood through the Function/Expression theme.[8] For example, similar gestures performed in different reach spaces can have very different purposes, and therefore different meanings. Consider a hand waving back and forth with an open, relaxed palm facing away from the mover. When made close to the body, in near-reach space, contacting the nose, the act might be to the functional end of scratching an itch on the nose—and similarly interpreted by onlookers as an expression that the mover has an itch on the nose. In mid-reach space, the action might be a gentle wave to serve the purpose of signaling to a nearby friend—and seen as a friendly expression of "hello" by onlookers. In far-reach space, the action might be used to ward off unwanted attention—and seen as, quite opposite to the previous example, a signal to "go away." Thus, the same movement performed from a different aspect of the kinesphere will influence the meaning of that movement.

The relationship between function and expression is used by dancers training to complete complex tasks with functional efficiency and economy, expanding their personal palette of available movements and, thereby the ideas that they can express with their bodies. In other words, by becoming functional, efficient movers, they become more expressive artists. The inverse is true as well. Factory artisans who complete complex, nuanced, and skilled physical labor employ a wide range of movement styles and qualities of expression to execute tasks like polishing, painting, and assembling. In other words, a broad palette of available actions allows them to complete concrete, functional tasks in their workplace.

The Function/Expression theme can be applied to any component of the system, where it is constantly revealing the indivisibility of these ideas. Developing varieties of salient movement profiles for robots is often described as "bringing expressiveness to movement," but this framing is in conflict with this fundamental duality. In fact, all movement is both expressive and functional, but some movement *systems* (be it a particular dancer or athlete, or even one robot versus another) can be more expressive than others—which also makes them more functional tools. In discussing function and expression in the BESST System, we therefore often talk about *foregrounding* one concept over the other. We can certainly work on developing a new movement pattern or robot to an expressive (or functional) end, but we must always recognize the relationship expression has to function (and vice versa).

Embodied Exercises

- **Choreography and meaning:** This exercise uses the BESST taxonomy to generate movement and your imagination to invent uses for that movement.
  - Develop a movement sequence through changing levels: start from lying on the floor, then move to sitting, and then to standing.
    - Come up with a reason for the movement: invent a context, situation, and environment where the sequence takes on meaning.
      - Try one meaning that foregrounds the function of the movement (e.g., lying on the floor for rest).
      - Try one meaning that foregrounds the expression of the movement (e.g., standing up for oneself against a bully).
  - Try the sequence by only moving in near-reach space, then mid-reach space, and then far-reach space.
    - How do the potential expression and function, identified in the previous step, change for each iteration of the same sequence in distinct reach spaces?
  - Compare and contrast the functional efficiency and expressive capacity of the sequence as it morphs through other terms we have learned so far.
- **Observation and analysis:** This exercise uses the taxonomy and the Function/Expression theme to analyze movements found in your environment.
  - Go to an area that you frequent that is good for people watching. Identify changes in elements introduced in this chapter.
    - Come up with a reason (or meaning) for the movement you see, observing context, situation, and environment where the sequence takes place.

- Try one meaning that foregrounds the function of the movement.
- Try one meaning that foregrounds the expression of the movement.

Chapter Summary

This chapter has described several abstractions for organizing human movement (and thus human perception of nonhuman movement) in space. We have introduced the idea of the kinesphere and shared dissections of space ranging from refined, pointlike pulls to broad, swathlike areas that help describe movement—across distinct bodies—in their kinesphere. We have also described the practice and some of the theory behind movement scales, listing these scales in appendix B for the reader's further exploration. Then, a review of a project in movement specification showed how these concepts can be applied to creation with machines. Finally, we visited the Space component of the BESST System through the theme of Function/Expression, allowing some integrated examples and opportunities to absorb this larger principle. We learned that the organization of animate bodies in space is not simply a task of measurement (which traditional tools in engineering, design, and robotics are well equipped to accomplish), but also one of broad perceptual strokes that are applied in a relative way to physical bodies in their environment.

6 When Is the Movement Happening? The Temporal Perception of Movement (Time)

Chapters 4 and 5 have examined elements of answering "what" and "where" questions about movement. By establishing a rich set of descriptions for moving bodies and the space in which they move, we outlined the Body and Space components. In this chapter, we are going to answer questions related to "when," completing a basic triad of concepts that describe bodies moving in space and time. That is, in order to perceive change, and thus in order for movement to occur, we need a body moving in space *and* time; thus, we now consider the Time component as introduced in (LaViers & Maguire, 2022). Time is foundational to the experience of change, and thus the perception of movement. In these three components, we find the foundational bits of movement that even basic, ordinary machines can potentially generate and interpret. More complex ideas like quality, intent, and meaning will be covered in later chapters, which will further complicate the kinds of patterns that people perceive about (and enact on) bodies in space and time.

In explicating a fifth component,[1] we are expanding on the concepts traditionally covered as "phrasing" and creating a needed taxonomy for discussing the temporal aspect of movement. Humans today have an increasingly prominent and quantitative relationship with time: devices track user screen time, measure the length of exercise activities, and predict when passengers will arrive at their destination, updating them with estimates about how disturbances like traffic will affect their arrival time. Moreover, in the movement arts, choreographers have access to video recorders, can create lighting cues with precise timing, and fit their work into increasingly varied event durations.

It is especially useful to explicate the Time component when working with machines, which must deal in user-specified and -designed quantitative units of time. Some machines do not portray clear ideas about intent and relationship from their movement, but we can always measure the amount of time an action took and frequently see phrasing through stops and starts of various machine parts. Thus, while a particular example of artificial movement may not rise to the level of creating a clear dynamic quality (which we discuss in chapter 8 about the Effort component), it will use elements that we are collecting in this component of Time. For example, an acceleration of a distal linkage of a machine, which creates a whiplike snapping action, may or may not manifest as a recognizable shift in movement quality, as reading significance in such an action greatly depends on the features of the movement preceding and following it, as well as other contextual factors. Likewise, algorithms interpreting human movement do not always recognize motivation or intent but can often segment motion into salient pieces.

Creating machines that move (or interpret movement) with variations in time that are meaningful to human observers is a crucial challenge. For example, pedestrians at crosswalks observe the oncoming cars and get some sense of how long they have until the cars reach the intersection. They also interpret temporally meaningful actions taken by the driver: if the driver speeds up, for example, a pedestrian will not cross. However, even if the car slows such that the actual time-to-arrival allows the pedestrian safe passage across the intersection, the pedestrian still will not cross if the driver's braking does not clearly advertise an intent to yield.

As this example of pedestrian and driver suggests, there are quantitative and qualitative aspects to time that we will discuss as part of the Time component and are critical to meaning-making. Moreover, the driver and the pedestrian each have different prior experiences, personal preferences, and goals in the context of their interaction; thus, they have distinct relationships to time. In general, a mover's relationship to time is subjective and personal, as expressed through an idea called "dynamosphere," which is introduced in this chapter.

6.1 Sequencing
Creating patterns through temporal context

The order in which actions occur and unfold over time affects the perceived meaning of the movement expression. What comes first provides context to what follows, and what follows reframes what just happened, influencing where and which patterns are evident. The **sequencing** of movement events affects the perceived emphasis—and therefore the experience and interpretation of the event. What happens first, what follows, and what comes after that creates a sequence that can be viewed in terms of a beginning, middle, and end. When the sequence of what happens first and what happens next changes (but the same set of movements is kept), the meaning of the movement will sometimes change as well. For example, consider the marshaller who directs planes on the ground at airports. This person uses a set of predesigned movements that indicate to pilots how to drive airplanes in particular ways. The sequence in which these gestures are performed becomes critical to the movement response of the aircraft pilot. To get the plane safely away from the gate, the marshaller will signal to back up, and then turn, and then go forward. If the turn signal is given before the backup signal, the plane will hit another plane at an adjacent gate, so the task will not be effectively completed. The sequence of baton gestures will indicate a particular meaning in one order, and if the same gestures are done in a different order, the meaning is changed and the outcome will be different. Even though the same movements are being performed, a different order creates a new meaning.

In observing or experiencing movement in sequence, we can further differentiate whether more than one thing is happening at the same time. When more than one thing is occurring at the same time, we say that these events are using **simultaneous sequencing** (i.e., in parallel). When individual actions follow upon one another, these events are using **nonsimultaneous sequencing** (i.e., in series). In the example of directing a plane on a tarmac, the "backup" baton gesture is done with the right and left arm moving at the same time, with the arms parallel to each other, and the "turn" baton gesture is done by putting both arms up overhead and *then* moving the other arm in the desired turn direction.

Nonsimultaneous events include the other two types of body-level phrasing first described in box 4.5 in chapter 4: **successive sequencing** (adjacent body parts following a movement expression) and **sequential sequencing** (nonadjacent body parts following in a movement expression). The observation and experience of successive and sequential body-level phrasing as nonsimultaneous sequencing will rely on body morphology. Thus, we articulate an affinity between concepts in the Body and Time components. The adjacency of body parts is different for a body with two limbs protruding from the core than for a body with forty-nine limbs. The idea of adjacent and nonadjacent body parts—what those parts are and how they move in concert—is determined by the individual context of each body's form. For example, the successive movements of a jellyfish and its waving tentacles moving from the pulse of its bell opening and closing are quite different from those of the cheerleader who makes a series of sequential gestures to complete a "Go, team, go!" chant. Likewise, robots that are just an "arm" and a "hand" will have different expressions of these ideas than those that have a "head," "trunk," two "arms," and two "legs." We suggest that movement phenomena are easily perceived as either at the same time or not at the same time and, given our propensity for finding and seeing patterns, sometimes things that happen at different times look behind or ahead with respect to one another.

There are choreographic techniques that manipulate how movement unfolds over time across multiple bodies using more complex aggregates of these basic sequencing ideas. Such tools include **unison, out-of-step, repetition, reversal, retrograde, accumulation, canon, echoing,** and **theme and variation.** These structures, listed in box 6.1, describe manipulations of how many times and in what sequence movement events occur, using sequencing to create pattern and emphasis.

In some of the sequencing strategies listed in box 6.1, emphasis on a particular moment within a movement phrase or expression is created through contrast. That is, emphasis can emerge when a sequence of actions is similar and then a very different type of action occurs relative to the movement events before and after. The moment of difference will often stand out, although crucially, not every moment of contrast is perceived in the same way by every viewer. For example, the airplane marshaller may see an unexpected change in the environment, causing the need for a pilot to quickly change course. The marshaller may then, after executing several evenly timed and similarly moved actions, quickly tense his arms and bring them overhead rapidly, indicating that the pilot should stop immediately. This change in movement tone or quality signals new information to the pilot through contrast.

Some mechanisms used to create contrast, and thus, for some observers, emphasis include *changing movement complexity, condensing action, muscular tension, spatial arrangement, contact,* and *sounding.* These structures, listed in

---

Box 6.1

Choreographic Terms for Sequencing Strategies

To illuminate the differences in these ordering structures, each description uses the following set of basic body actions: expand, condense, rotate, jump, and locomote.

- **Unison:** Performers do the "same" series of actions at the same time. This is a baseline case, from which these other strategies may be seen as variations.

- **Out-of-step:** Each mover again performs the same sequence, but not exactly at the same time as every other mover.

- **Repetition:** Any action or set of actions that is repeated more frequently inside a longer sequence will emerge as significant to the understanding and emphasis of that moment in the meaning of the whole sequence. Regardless of whether these actions are repeated in succession or across a longer sequence where they are mixed with other actions, but repeated noticeably more times than all the other actions, the effect is an emphasis on that action or set of actions as being more prominent than the other actions in the sequence. So, if the sequence is performed as *expand, expand, expand, condense, rotate, jump, locomote,* the action of "expand" is emphasized. Likewise, if the sequence is performed as *expand, condense, expand, rotate, expand, jump, expand, locomote,* "expand" again becomes the emphasized action. At some point, after many repetitions, a process of desensitization can also occur.

- **Reversal:** The original sequence would be ordered as *locomote, jump, rotate, condense, expand.* Thus, the same set of actions performed in reverse order changes each movement in the sequence, as each action needs to connect to a different action than in the original sequence.

- **Retrograde:** This can be understood as akin to putting a video in rewind mode. The order is reversed, but the movements are also performed by tracing backward the actions that occurred going forward in the sequence. Enacting this structure will create a different sequence of actions altogether, but each action within the sequence preserves its original form (just in "rewind").

- **Accumulation:** This way of ordering movement relies on repeating the previous set of actions and then adding a new action. Using the previous example set of actions, a mover creates a sequence of *expand* (1), *expand, condense* (2), *expand, condense, rotate* (3), *expand, condense, rotate, jump* (4), *expand, condense, rotate, jump, locomote* (5). By the fifth repetition, the entire phrase is revealed.

(continued)

Box 6.1 (continued)

- **Canon:** Two movers both use the same sequence of actions, but shifted in time. The first mover performs *expand, condense, rotate, jump,* and *locomote,* while the second mover waits (often a measure of music) to perform *expand, condense, rotate, jump,* and *locomote.* This type of ordering is also seen quite clearly in singing a "round" such as "Row, row, row your boat (1) gently down the stream (2), merrily, merrily, merrily, merrily (3), life is but a dream (4). The first singer would begin with "Row, row, row your boat" and continue through the sequence. The second singer begins the sequence after the first singer has completed the first set of words. However, both sing at the same time and follow the same sequence, just shifted in time.

- **Echoing:** One mover performs *expand,* then the second mover performs *expand* as an "echo" to the first mover, and so on through the sequence. This strategy uses the same time shift as canon, but not the same accumulation.

- **Theme and variation:** Riffing on a movement theme, creating a slight variation in the form of a new movement, allows a larger multiplicity of sequencing choices. Some examples include:

  ○ ABA ("theme" is A, "variation" of the theme is B)

  ○ ABACADAE (C and D are further variations of A)

  ○ ABCABDABE (the theme can now be seen as AB in this sequence)

  ○ ABABCABABDABABE (the theme is either ABAB or a repetition of AB, interrupted with C, D, and E)

box 6.2, describe manipulations of how movement is presented over time, using contrast to create pattern and emphasis within a temporal stream of actions.

By noticing the relationships of actions in time with greater detail, we open up new options for design and interpretation. Sequencing (the order of actions) creates contrast (a measure of action relative to temporally adjacent actions) and emphasis (the value, prominence, or importance of actions), creating a temporal pattern and context. This contrast or emphasis can be created through an *abrupt* (or *gradual*) change that stands out among the actions temporally close to it. An abrupt action seems like a dramatic moment of intense change, while a gradual action seems to unfold slowly, with less change happening in any given moment.

Box 6.2

Choreographic Terms for Creating Contrast

- **Increased (loading), or decreased (unloading), movement complexity:** Actions that dramatically change in their complexity may be perceived as being in contrast to those around them. This could happen through many lenses of movement. For example, using a central pathway is often seen as a simpler spatial choice than a transverse pathway through the kinesphere. So a series of central pathways could be seen in contrast to a transverse pathway, which would be emphasized through its distinct sense of complexity; or vice versa, a central pathway among many transverse pathways could be emphasized through its simplicity. Often, this idea is called *loading* or *unloading*, respectively.

    ◦ **Stillness:** An important subcase of this idea is going between active movement and stillness in the body such that a viewer perceives stillness (simplicity) versus action (complexity).

- **Condensing dynamic expression:** Actions that happen with force and impact, speed, piercing focus, and extreme binding (see also the discussion of condensing effort qualities in chapter 8) draw attention and create emphasis. This idea is also sometimes called an *accent*, especially in word pronunciation (which is a subset of movement behavior), where this is sometimes understood as the loudest syllable of the word when spoken. Contrast can also be created through a moment of softness and relaxation amid stronger action.

- **Muscular tension:** Actions that require extreme change in the attitude of the musculature (especially contraction, but also relaxation) can create emphasis.

- **Spatial arrangement:** The place within the environment where an action unfolds can create emphasis. A sequence that goes toward the audience compared to the same sequence moving away from the audience will create a different impact. For example, according to the choreographer Doris Humphrey, the diagonal from left upstage to right downstage was considered the strongest diagonal on the stage, and movement on that pathway would have the greatest emphasis (Humphrey & Pollack, 1959). This may have been true for Humphrey's style of movement on a traditional proscenium stage, but every specific environment has a spatial structure with which movement can engage for particular emphasis.

- **Contact with self/other/the environment:** Actions that involve touch or contact of some kind can emphasize the impact of a sequence. Take, for example, a mover that begins with an exaggerated clap (expand, condense),

(*continued*)

Box 6.2 (continued)

opening the arms and then clasping the hands together, or does a "high five" with another mover, or slaps an object in the space. These moments of contact draw importance to the "condense" event inside a larger sequence of movement. Or, in contrast, in a series of movements that all use contact, an action that disengages contact from the environment will be emphasized.

- **Sounding:** Using percussive action that creates sound, such as a slap or the voice (the basic body action of vocalizing), can draw attention inside an otherwise silent movement. The moment of sounding stands apart as different from the other moments in a sequence that does not involve using the voice. Or, in contrast, in a series of movements that all use vocalization, an action conducted in silence will stand apart from them.

This forms an affinity between the Space and Time components. In section 5.3 in chapter 5, we described abrupt and gradual *spatial* changes occurring in movement scales. Here, we recognize that an abrupt spatial change that happens in a short period of time will seem even more abrupt than one that occurs over a longer period. This is an example of the Space and Time components reinforcing—or heightening—expression in movement.

Embodied Exercises

- **Transforming movement through time:** This exercise invites you to consider the physical and interpretive implications of order.

  ◦ Make a short movement sequence that starts by lying on the floor, goes to sitting, and finally to standing (as in the first embodied exercise of section 5.5 in chapter 5).

  ◦ Now reorder the sequence from sitting to standing to lying on the floor, and then from standing to lying on the floor to sitting.

  ◦ How many ways can you order these three actions? What changes when you do each combination? Consider the answers to this question through the lens of Body, Space, and Time, as well as possible interpretations of motivation and intent.

- **Reordering beginning/middle/end:** This exercise invites you to reorder and repeat (forming emphasis) gestures that may create a different narrative depending on how they are structured in time.

- ○ What is happening when you point, wave, and beckon? Is it different if you beckon, wave, and point? Or if you wave, beckon, and point?
- ○ Now choose one of these actions to repeat; for example: point, point, point, wave, beckon.
- ○ What does the sequence reveal now? How does the sequence of events and repetition change the possible meanings of the movement event?

- **Changing the sequence of emphasis:** This exercise uses a familiar inroad to movement (vocalization) to explore emphasis.
  - ○ Try speaking the words, "Who am I."
  - ○ Now try emphasizing the first word (e.g., with increased volume compared to the other two words) "*Who* am I"; then the second word, "Who *am* I;" and finally the third word, "Who am *I.*"
  - ○ How does the meaning and experience of your expression change based on which word is emphasized? Is one version a statement and one a question? Note your experience and responses to each version to begin to understand the role that emphasis plays in meaning-making.
  - ○ Try recreating the "same" phrase without speaking. How do you create emphasis in your body without using your vocal cords?

6.2 Duration, Tempo, and Rhythm
Marking time and our perception of it

Our perception of movement is affected by **duration** in multiple ways. It changes our sense of whether a mover was moving "fast" or "slow" (both relatively and absolutely); it affects our estimate of the mover's intent and inner status; it often governs whether an action is judged to be salient and complete or unintentional and errant. The limit of human perception of duration occurs at the scale of milliseconds and has been measured by scientists in experiments with human subjects. This limit is generally accepted to be 100 milliseconds: any event that occurs at this rough threshold is typically judged to be instantaneous (or, events that occur in less than ~100 milliseconds typically go unnoticed by the human viewer). This is an important threshold that probably evolved based on the types of activities that humans typically engage in: most activities do not require perception below this threshold; however, this is a domain where machines, like high-speed cameras and precision lasers, aid—and help quantify—our natural abilities. Thus, we first establish a taxonomy of *relative duration* and *absolute* *duration* to describe measurable features of time. This taxonomy is given in box 6.3. Later in this section, we introduce a notion of quality in order to further distinguish what we can experience from what we can measure.

Actions of different durations may be more salient for different application areas. Here, we rely on existing, established measures of time to create a taxonomy of activity levels[2] to provide descriptive power enabling statements like "a contraction of the pointer finger that lasted with *twitch level duration*, which was *shorter than* that of the overall expansion of the entire hand." In this case, these terms help us resolve the temporal scale of different bodily actions, indicating that a small contraction of a finger does not negate the overall expansion of the entire body part of the hand. In this book, we consider duration to be interchangeable with **speed;** that is, the same action done over different durations will have different speeds. (We do not explicitly consider velocity vectors, as these are well understood inside traditional robotics texts.)

Box 6.3
Relative and Absolute Duration

- Relative duration
  - Short or shorter than
  - Equal to or the same as
  - Long or longer than
- Absolute duration
  - Micro (patterns in movement; the focus of this book)
    - Instantaneous (less than 100 milliseconds)
    - Twitch level (tenths of a second)
    - Action level (seconds)
    - Phrase level (minutes)
    - Activity level (hours)
  - Macro (broader patterns in behavior)
    - Days
    - Weeks
    - Months
    - Years

As soon as we begin considering duration, we can think about temporal patterns in duration, specifically **tempo** (as measured, for example, by beats per minute) and **rhythm** (as described, for example, by **meter**). Rhythm is a way of marking time and is inherent in our experience of our bodies moving in the world. Our biological functioning is based on rhythmic patterns that coordinate different body parts for the purpose of successful movement. Take, for example, the fundamental expression of our heartbeat. Ideally, it pumps with a steady, duple (two-phase) rhythm, but when the beating of the heart changes, it is called "arrhythmia" (out of rhythm). Our breath also functions on a duple rhythm (inhale/exhale), but that rhythm can also change and be quite impactful on our experience. Take, for example, the statements "I can't catch my breath" and "I'm out of breath." These refer to experiencing the breath in different or unusual rhythmic relationships.

In the BESST System, we identify certain skeletal/muscular rhythms to enhance and optimize our range of movement and clarity of intent. These ideas relate the movement of body parts to temporal patterns. Moreover, musical structures such as three-four (or three-quarter) time also provide known temporal patterns that can be used to modify the practice of movement. Common rhythms discussed in movement studies are both internally and externally manifest, described in box 6.4.

In addition to specific rhythms, we are concerned with the qualitative experience of time, which is not always coherent with the measurable passage of time (in absolute or relative terms). For example, anticipation modifies our experience of time: waiting for the result of a job interview can *feel* endless, even if it is only a few days; on the other hand, watching children grow up can *feel* instantaneous, even though it takes years. Moreover, the way that bodies move around us changes our experience of the passage of time. For example, a room filled with people running timed agility drills and a room filled with people soaking in a hot tub have different aesthetic qualities, including that time seems to pass differently in each. Consequently, we draw a contrast between duration and perceived duration, which we call **time quality,** and enumerate some options of how the subjective experience of time may be described (or how time *feels)*. We organize these ideas as pairs along the dimensions of *gradual* and *abrupt* poles in box 6.5.

These qualities are modified by other elements of the Time component as well. For example, an event lasting 5 seconds can seem to have *an instantaneous time quality* if preceded by an event lasting 5 minutes, or seem to be

Box 6.4
Rhythmic Structures

Absolute

- Tempo[3]
  - Measured in beats per minute
- Scheduled events
  - Measured in years, months, days, hours, minutes, seconds, etc.
    - For example, Halloween creates an annual collective increase in candy consumption in households in the US.

Relative

- Internal (coordination of different body parts for successful movement)
  - Breath, heartbeat
  - Muscular/skeletal rhythms
    - Occipital/sacral rhythm (between the occipital bone and the sacrum, as in heel rock and knee drop)
    - Gleno/humeral rhythm (between the glenoid fossa of the scapula and head of the humerus, as in arm circle)
    - Ilio/femoral rhythm (between the base of the ilium and head of the femur, as in thigh lift)
  - Ultradian (hourly), circadian (daily), infradian (monthly) rhythms
- External (coordination of different bodies for successful movement)
  - Synchronization through rhythm
    - For example: dancers, rowers, people lifting a heavy object, switchboard operators, and assembly line workers use audible counting to coordinate action.
  - Rhythmic structures (meter)[4]
    - Three-four time
    - Four-four time
    - Six-eight time
    - Duplets
    - Triplets
    - Iambic pentameter
    - Dactylic hexameter

Box 6.5
Time Quality

* Gradual polarity
  ◦ Lingering
  ◦ Prolonged
  ◦ Ongoing
  ◦ Endless
* Abrupt polarity
  ◦ Instantaneous
  ◦ Immediate
  ◦ Rapid
  ◦ Stopped

---

prolonged if preceded by an event lasting half a second. Thus, the sequencing of events of different durations changes their perceived time qualities. Likewise, a movement event that is emphasized through some means may seem to last longer, as we are attending to it in more detail, sensing some importance in it, than the surrounding events in the sequence.

Embodied Exercises

* **How long is 1 minute?** This exercise will expose our qualitative relationship to time through one of our readily accessible tools for measuring time.
  ◦ Get a stopwatch (or stopwatch app) handy.
  ◦ Start the timer and, as you do, close your eyes.
  ◦ Endeavor to wait exactly 1 minute.
  ◦ When you think that 1 minute has passed, open your eyes. Did you overshoot or undershoot?
  ◦ Try again. Did you get closer?
* **Relating duration and tempo to the body:** This exercise aims to be the reverse of the previous exercise, revealing how our movement patterns—in this case, we will explore breath and tempo—inherently possess quantity.

◦ Find a pair of songs with different tempos (an upbeat, fast option and a slower-paced choice). Repeat the two experiences below for both pieces of music.

- Use an even breathing pattern: inhale for four counts and exhale for four counts.

- Use an uneven breathing pattern: inhale for three counts and exhale for six counts; and then reverse this relationship.

◦ Reflect: How does experience change as you change the duration and tempo of each action? Which ratio of inhaling and exhaling felt most comfortable or familiar? Which tempo felt most comfortable for timing your actions?

• **How long do you think an activity will take?** Estimating time is part of our daily budgeting process. Check your estimate the next time you engage in a relatively unknown task (like reading one page of this book). How long did you think it would take? How long did it actually take? How long did it *feel* like it took? Which words from the list of time qualities provided here fit that activity best?

6.3 Phrasing
Patterns in time that manifest coherent and individual movements

In movement theory (as in music theory), **phrasing** refers to the idea of grouping movements (or notes) that belong together in a single temporal stream. These groupings affect how we understand a movement activity, answering the questions "How many things happened?" and "What was the relationship between these things?" A phrase is often thought of as having an initiation (beginning),[5] a main action (middle), and a conclusion or resolution (end). As phrases are perceived through the passage of time, the ideas discussed in the previous sections, like duration, begin to have an inherent impact on the perception of a phrase. Emphasis, in particular, is an important component of how we perceive phrases. For example, creating a sense of emphasis through rarified dynamic quality (a punch at the end of a series of softer, slower movements) can punctuate the end of one phrase and signal the beginning of the next.

The idea of phrasing exists throughout the movement components of the BESST System: in Body (for body-level phrasing, see chapter 4), in Space (for volute and steeple phrasing, see appendix B), and in Effort (for dynamic organization, see chapter 8). Often the interaction of these components produces a sense of phrasing. For example, seeing a series of movements sequencing along adjacent body parts (as in successive body-level phrasing) tends to group those actions as being related to the same phrase.

Phrases have malleable temporal scales across which one can exist. For example, a day can be perceived as a phrase of time over 24 hours and complete unto itself, but days are also parts of the longer phrase of a week. We can observe movement and activity for a day, but we can also observe the pattern that occurs over a week.

The same is true for any bodily movement: a given action can be seen as an event itself or a smaller piece of a longer, more complex task. Thus, phrasing is the way that the BESST System formalizes the idea of "a movement" compared to "a series of movements." What makes "a movement"? What makes a phrase feel like "four movements" versus "five movements"? The answer is contextual. Moreover, our sense of temporal phrasing interacts with our bodily, spatial, dynamic, and relational senses of phrasing. Imagine a piano player. We see the hands and fingers engaged in pressing keys with rapid succession to produce notes, but we also can see the feet engaging with the pedals on a different timescale, sustaining and dampening different notes for a desired effect. Although the whole body is moving the whole time, we can either see this as two simultaneous temporal phrases or one longer, more complex whole-body phrase. If the piano pedals were instead keys on the piano, which would constitute a new environmental context, we might perceive this differently due to the different body organization that is producing the sounds.

In chapter 5, the spatial phrase of moving from right-forward-high to left-back-low was an example of abrupt change in *space*. All three spatial pulls change from the beginning of the phrase to the end. Likewise, moving from right-forward-high to right-back-high is an example of gradual change, as only one spatial pull changes from the beginning of the phrase to the end. Here we introduce the notion that temporal phrases can be gradual or abrupt based on both their duration (as a phrase, which can be thought of as a single movement or multiple movements) and their time quality. Explicating temporal and spatial aspects of phrasing allows us to differentiate between temporal and spatial change as well as contextual features that impact how long an action *seemed* to take.

For example, the spatial phrase between right-forward-high and left-back-low could happen with a long duration, lessening the sense of abruptness overall. What we can say, then, is that the phrase has abrupt spatial change because it is moving between drastically different points in space, and gradual temporal change because it takes a long time. But if the change in spatial pulls occurs between actions of even longer duration, this context may produce an accelerating time quality for the phrase.

Thus, choices in temporal organization of movement change how we organize, experience, and perceive meaning based on where in the phrase the emphasis occurs. Naming the types of phrases and the relationship between them (as we do next) helps us tease apart (or analyze) and make sense of (or synthesize) movement in the temporal dimension. Box 6.6 and figure 6.1 present a list of and symbols for several types of phrases that reveal different patterns of accent, loading, and emphasis.

Identifying types of phrases and their character or quality based on emphasis, loading, and accent allows us to analyze phrases as a larger whole with meaning and organization rather than just a series of actions unfolding over time. The same set of movements phrased differently will change the meaning and experience of each movement, as well as the set of movements overall. Take, for example, the following set of actions:

- Get up from a chair at the kitchen table and walk over to the stove.
- Turn off the burner under a pot.
- Sit back down.
- Tap one foot.

Box 6.6
Types of Phrases

- **Even:** All actions have the same emphasis (without accent).
- **Impulsive:** The beginning actions are accented (loaded or emphasized).
- **Impactive:** The ending actions are accented (loaded or emphasized).[6]
- **Swing:** The middle actions are accented (loaded or emphasized).
- **Becoming (or increasing):** The actions build in emphasis over the course of the phrase (analogous to a crescendo in music).
- **Diminishing (or decreasing):** The actions become less emphasized over the course of the phrase (analogous to a decrescendo in music).
- **Vibratory:** The actions in the phrase rhythmically spike and wane, often with a fast tempo, but over time, they can appear to have the same value.[7]

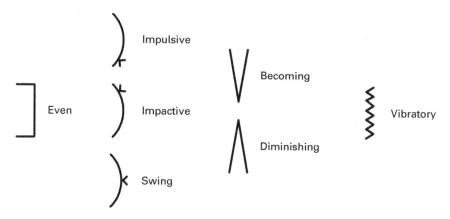

Figure 6.1
Symbols for types of phrasing. These "phrasing bows" wrap around other symbols
that indicate action to denote the clustering of movements into distinct units.

If one of these actions is emphasized (say, the first action happens quickly,
with increased muscular tension), as opposed to all the actions in the
sequence being given the same relative duration and muscular tension over
time (resulting, say, in even phrasing), a different meaning emerges. This
choice of phrasing could be seen as simply attending to something that is
cooking, while the impulsive phrasing of emphasizing the first action could
indicate something urgent occurring (e.g., the pot boiling over) and a need
for intervention to forestall a mess or other calamity. A constant (simultaneous) tapping of the foot in parallel to the other actions may be seen as
an example of vibratory phrasing, creating a sense of nervous energy about
the subject who is cooking.

Phrases can also be seen in relationship to other phrases as the action
unfolds over time. Types of arrangements between phrases include those
listed in box 6.7.

An example of consecutive phrasing can be seen in opening a jar lid.
First, the jar is grasped. Then, the hand is placed on the lid and turned to
loosen it. Finally, the lid is removed from the jar. Each phrase of movements completes before the next phrase begins. When someone juggles
multiple objects, it is often possible to see how he is tossing with one arm
while catching with the other before the toss has finished and preparing
to toss again in the middle of a catch; we can name tossing and catching
as two distinct phrases that alternate on each arm with an overlapping
relationship. Simultaneous phrasing can be observed when the mover is

---

Box 6.7
Relationships between Phrases

◦ **Consecutive:** The phrases unfold one after the other, as each completes before the next begins
◦ **Simultaneous:** Two phrases happening at the same time
◦ **Overlapping:** A new phrase beginning as the previous phrase is ending

---

doing a set of actions with the arms while doing a different set of actions with the legs at the same time, as in the previously described example of the piano player (hands versus feet). Thus, the relationship between phrases adds further descriptive power to discerning the temporal structure of movement.

Finally, as we have described in this chapter, our experience of time has a qualitative, subjective component. This is captured in the idea of a **dynamosphere.** Just as a moving body has a *kinesphere*—a physical space where its movement is perceived—a moving body can be thought of as having a temporal "space" where its movement is perceived. This temporal component of our movement is often associated with intent: if something is very important to us, we may choose to do it first, whereas if it is not important, we may do other things first. This simple example relating sequence to dynamosphere reveals how personal our choices in time are. The dynamosphere becomes especially rich and complex when quality (see chapter 8) is considered as well, forming an affinity between the Time and Effort components.

Just as we perceive space around familiar moving bodies as capable of being filled with their presence (a concept captured by the kinesphere), we also perceive bodies as being able to make meaningful, purposeful, and personal choices in time (a concept captured by the dynamosphere). These dual concepts of kinesphere and dynamosphere parallel, to an extent, the use of the terms "kinematics" and "dynamics" in engineering: the kinesphere deals with the ability to move the articulated body (often thought of anatomically as our joints) in space, while the dynamosphere deals with the ability to move the weighted, physical body (often thought of anatomically as being created by our muscles) in time. However, the kinesphere and dynamosphere emphasize the personal experience of a mover rather than the universal experience of objects acting under gravity that kinematics and dynamics explain.

Embodied Exercises

- **Using vocalization to experience phrasing and its effect on meaning:** Using this sequence of four words, "I am here now," try the following and how it changes the meaning of their expression.
  - Begin by stating the four words with equal emphasis (e.g., same duration and volume): "I am here now" (even phrasing).
  - Emphasize the first word: "*I* am here now" (impulsive phrasing).
  - Emphasize the second and third word: "I *am here* now" (swing phrasing).
  - Emphasize the fourth word: "I am here *now*" (impactive phrasing).
  - How does the meaning and story change depending on where the emphasis is?
  - Try vibratory, becoming, and diminishing phrasing for this sentence as well.
- **Explore spatial and temporal phrasing with gradual/abrupt change:** Revisit the spatial pulls in the icosahedron (as described in chapter 5) to explore different phrasing:
  - Move the actions described here (sequence A):
    - Using your right arm, move from left-side-high to forward-low to right-back-middle as one phrase.
    - Continue on from right-back-middle to left-side-low to forward-high as a second phrase.
  - Move the actions described here (sequence B):
    - Now start from right-forward-middle and go to left-side-high, ending in forward-low, moving this as one phrase.
    - Continue moving the right arm from forward-low to right-back-middle, ending left-side-low.
  - Note the common points of spatial direction (left-side-high, forward-low, and right-back-middle are reused) between the two sequences.
  - How many movements are in each sequence? Does each bullet point feel like one movement or two?

◦   How do these distinct choices in spatial phrasing change your temporal performance of the scale? Do you perform sequence A more slowly than sequence B? If not, try this out. How does this support a sense of gradual change in sequence A and a sense of abrupt change in sequence B?

These two sequences play with volute (sequence A) and steeple phrasing (sequence B). Typically, the two phrases in sequence A feel like one movement each, while the two phrases in sequence B feel like two movements each. See appendix B for more about these types of spatial phrasing.

6.4   Application to Machines: Perceived versus Actual Safety
Choreographing the motion of aerial vehicles

When programming the temporal patterns of robots, quantities like "time to collision" calculate how soon under given applied forces a known mass will collide with another (e.g., an aerial robot and an obstacle in its environment). In a traditional factory, where humans are typically separated from robots by safety measures like cages and other physical boundaries, such a calculation is enough to facilitate successful operations. Psychologists have studied a similar quantity in human perception, aiming to estimate how people judge such a "time to collision," but in humans, such a benchmark carries both functional and expressive dimensions. Roboticists we've worked with have reasoned that robots are socially acceptable if they do not move too fast (because fast robots are scary).

But this is an example of the pitfalls of working with an incomplete framework for understanding expression. It is as easy to imagine a counterexample: for a really, really slow-moving aerial robot, it will be difficult to predict where it is going next, which becomes annoying—and eventually even creepy and unnerving. This is happening with autonomous factory robots that share spaces with human workers. Factory workers duck around these mysterious, monolithic devices that do not clearly advertise their internal state or broadcast their next movement with changes (including speeding up) to their temporal motion profile. Endlessly even phrasing patterns cause human attention to stray. This can create a dangerous situation—even if every quantifiable equation (e.g., the movement of the robot) says the opposite.

Consider the role of duration in the example referred to at the chapter's opening: a pedestrian steps into a crosswalk with cars still moving toward the intersection. Even if the drivers of the oncoming cars are in control of their vehicles at all times, the way that the drivers approach the intersection will inform the pedestrian's feeling of safety—situated in her own unique body and abilities—which will in turn inform her own behavior in the crosswalk. An oncoming car screeching to a halt feels like it is coming faster and will also arrive sooner, perhaps causing pedestrians to dodge or dash away, creating other safety concerns. Drivers use the motion of their cars to advertise to pedestrians that they will stop by using deceleration that is unnecessary from a functional point of view. Thus, the Time component is essential to finding somatic strategies for designing both aerial and terrestrial autonomous vehicles that share a space with humans.

Robots can advertise their internal state through both their physical motion (which is limited by the physical capabilities, or dynamics, of the platform) and dynamically unconstrained degrees of freedom, such as lights (e.g., light-emitting diodes mounted to the robot body that can flash on and off faster than physical linkages can move and even faster than humans can perceive). Figure 6.2 shows an example of both of these from a project to create assistive mobile devices for older Americans who are aging in place. The general idea for this work was to develop a system that relies on tunable parameters corresponding to the dimensions of effort (for

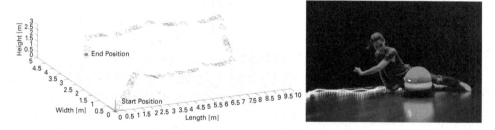

Figure 6.2

Using external changes to broadcast internal state. Left: modulating the movement of an unmanned aerial vehicle to create different textures, tones, and moods changes the duration and emphasis to the robot's motion (Cui et al., 2019). Right: adding lighted elements to a mobile robot creates different expressive modes, which can change the sense of time quality of the device's movement, as perceived by a human audience member (Pakrasi et al., 2018). Image by Keira Heu-Jwyn Chang, used with permission.

more about the Effort component, see chapter 8) and to produce motions of varying execution, which should cue human onlookers to the fact that the state of the system has changed. The system was to have specified way-points that were required for correct function (the "task"); but the motion between the points also had to reflect the context and internal state of the system. For example, movement patterns for a leisurely Sunday morning when the system is checking whether the paper has been delivered should be different from those used in an emergency, when the system is fetching important medicine that the user needs.

The result of our simulated artificial movement was a body that varied its velocity and trajectory, creating ten distinct patterns of aerial vehicle flight (Cui et al., 2019). The variations in the machine's motion created a sense of lift, or buoyancy, at times, and at other times, they created stark, sudden dips that gave a sense of jagged sharpness. Inside the white, gridded "simulation environment," these changes happened for no apparent reason, making it difficult to judge a sense of agency or purpose in the various flight patterns. However, across broad notions of expediency, efficiency, and overall success (reaching the end point without a crash), the profiles can be judged: a profile with a direct path of short duration can be seen as more expedient and efficient than one that meanders or weaves back and forth before reaching a final location in the simulated space.

While the work initially used the idea of changes in movement quality, it seems in retrospect that a better approach would have been to consider the phrasing of the flight trajectory of the aerial robot. Notions of temporal pattern changes in the flight path, organized as phrases of movement with various patterns of emphasis, loading, and accent, may well have been a more effective way to broadcast internal states to a human observer. As it was, each flight path seemed like one long, stumbling flow of words, lacking any punctuation. A simpler temporal structure, defined by clear differences in action in the beginning, middle, and end of each behavior, may have created a more readable distinction between each profile. Rather than aligning our tunable parameters with ideas of dynamic qualities, we could have aligned them with the simpler ideas of temporal qualities. The temporal changes could have then been used to support positive human-robot interaction (HRI) for in-home devices. For example, a slow, even flight path where the whole path showed no temporal fluctuation could well support the internal state of "monitoring" or "scanning" for security. Likewise, a flight path in which the initial speed was fast over a short duration (impulsive) might well have broadcast an internal state of "urgency," of being responsive to an urgent situation.

Embodied Exercises

- **How long is 1 minute? (Revisited with machines):** This exercise will explore how machines change our sense of time. Repeat the first exercise from section 6.2 with a blender (or food processor).
  - Get a stopwatch (or stopwatch app) handy.
  - With the machine on its highest/fastest setting:
    - Start the timer, paying attention to the machine's motion.
    - Endeavor to wait exactly 1 minute.
    - When you think 1 minute has passed. Did you overshoot or undershoot?
  - With the machine on its lowest/slowest setting:
    - Start the timer, paying attention to the machine's motion.
    - Endeavor to wait exactly 1 minute.
    - When you think 1 minute has passed. Did you overshoot or undershoot?
  - Qualitatively, did the two experiences feel different? Quantitatively, did you let the stopwatch progress for different amounts of time? Using the vocabulary introduced here, compare and contrast how the machine settings contributed to your perception of time.

6.5 Exploring the Themes: Time through the Lens of Exertion/Recuperation
Taking cycles of rest as an integral part of productivity

The theme of Exertion/Recuperation is another lens through which we see patterns of phrasing unfold over time.[8] We work and are awake during the day; we rest and sleep at night. What is exertive for one person (and how he recuperates) may be different for another person, and different in different contexts. Take, for example, an office worker who sits at a computer most of the day. For that individual, a nice jog after work can be a recuperative activity. For an Olympic track athlete, however, running is her "work" and involves great exertion, so a nice jog might not be recuperative for her at all! For this athlete, sitting at a desk and doing computer work in the evening might well be recuperative.

In this sense, it becomes difficult to classify one type of movement activity as *either* exertive or recuperative as a static quality because the classification depends on context. Thus, as context changes, what is exertive becomes recuperative, bringing back the image of the lemniscate, where one surface (or idea) becomes indistinguishable from the other. Regardless of the activity, there is a rhythmic pattern in the actions of exertion and recuperation. Overwork and a lack of adequate sleep and recuperative activity—typically, too long a duration of work with too short a duration of rest—lead to a lack of productivity; likewise, too much lethargy, boredom, and inactivity lead to a lack of relaxation. The balance of both over time is important to preventing injury, fatigue, boredom, stress, and burnout.

Indicators to counterparts about our energy levels help humans understand each other and work in better harmony. This idea has been adopted for machines, too. For example, Apple has employed an indicator light that has different states to mimic the "on," "off," "sleep," and "charging" states of their computers. A gently pulsing light (probably meant to imitate breathing) indicates that the computer is neither on nor off—recharging, but ready.

In the BESST System, there is a pursuit of balance that is reflected not only in the themes, but also in the entire system itself.[9] The theme of Exertion/Recuperation also speaks to the idea of balance, and as it is so clearly seen in the passage of time, it can be understood as a dualistic relationship that is especially relevant to the Time component. With that said, we could view it through the lens of any of the system's other components.

Embodied Exercises

- **Manufacturing feelings of recuperation:** Commonly, we see actions like lifting weights as "exertive," but in some contexts, such an act feels "recuperative." In this exercise, you will explore the Exertion/Recuperation theme through repetition.
  - Pick an action that feels recuperative—maybe stretching your arms overhead or yawning.
  - Set a timer for 3 minutes and repeat that action until the timer ends.
  - Note when the action becomes taxing or exertive after you have repeated it so many times.

○ Observe the moment of recuperation that occurs when you get to *stop* doing that action that had felt so recuperative at first.

• **How long is 1 minute? (Revisited again):** Try some variations on this exercise from sections 6.2 and 6.4 (using a stopwatch to see if you can identify 1 minute of time) to explore how exertion (and recuperation) affect your perception of time.

○ Try the exercise looking at different stimuli (e.g., a video of rush hour traffic versus a video of Olympic sprinters).

○ Try the exercise at different times during your day (e.g., when you are rushing to complete another task or when you are waiting for water to boil). Often after rushing around in a hurry, you will be more likely to undershoot, whereas on a leisurely weekend afternoon, you may be more likely to overshoot.

Chapter Summary

This chapter has introduced vocabulary for better noticing patterns in and perceptions of movement over time. In addition, it has described the impact of order and relative magnitude of actions. We have discussed how changing the sequence creates contrast and emphasis, which modifies and helps create a sense of style, intention, narrative, and meaning. Ideas about dynamic change in time—namely duration, tempo, and rhythm—were presented as additional (but distinct) modifiers that further support one's ability to make and perceive patterns of movement in time. Like sequencing and contrast, these three ideas are interrelated: we need a notion of duration to set up a notion of tempo to set up a notion of rhythm. Likewise, duration and tempo may be revealed through the structure, design, and use of a particular rhythm.

The most important way we think about time in movement studies is in finding "movements" themselves. The idea that movement can be associated with discrete chunks of time relies on a notion of phrasing, which, as we have shown, relates to all the other components of the BESST System: Body, Space, Shape, and Effort, but lives primarily in the temporal dimension discussed here. Chapter 8 will further explicate how affinities between components of the system reveal patterns in our perception and experience of movement. Finally, we have provided examples of machine design, as well as the theme of Exertion/Recuperation, to further investigate the Time component.

7 For Whom Is the Movement? The Relationship between Mover and Environment (Shape)

In the previous three chapters, we have illuminated foundational aspects of movement (body, space, and time). The Shape component is, in our inverted triangle model of the BESST System, a higher-order idea of movement building on the prior three components (Body, Space, and Time). It is the connection between body and space that reveals the body's changing form (shape) in relationship to the environment. Thus, the Shape component necessarily speaks to "interaction and relationship" of some kind, and we say that it answers the question "For whom?" about movement.

In adult human movers, we typically see all the components of BESST simultaneously. In analysis, we learn to foreground or background different elements of the system to hone in on whatever the most salient or important feature is in a given context with a given analytical aim. For example, if using the BESST System to work with a client who is experiencing lower back pain, we may focus on the area of Body, foregrounding principles such as weight shift, dynamic alignment, and axis of length. But, if working to help a ballet dancer better express a particular role, we may move to an area like Shape to help them analyze how they are approaching their relationship to other dancers in the piece. However, we could also use Shape when working through therapy with the client in pain, asking her to repattern the way that she connects her lower body to a chair while sitting and typing at a keyboard. By the same token, we could use Body to support the dancer's execution of a complex lift with a partner. All the components of the BESST System help to create successful outcomes of movement intention and expression.

The interchangeability of the components of the BESST System can confound the process of seeing and understanding each element on its own.

Moreover, it may contribute to why the analysis of developing movers—that is, babies, toddlers, and young children—is so useful in the work. By looking at babies, we can see human movers who do not yet have cross-lateral patterning in their body organization (Body); who are not aware of their entire kinespheres (Space); who do not create intentional phrases in their movement (Time); and who do not shape their bodies to accommodate complex, three-dimensional interactions (Shape). Through their simplicity, babies provide a clarifying picture of complex patterns that form adult movement.

What about robots? Do artificial bodies always manifest all elements of this system? Like babies, robots are bodies that move in time and space, but they do not contain the same level of natural richness, complexity, and even meaning as adult movers. And, like babies, we find that there is sometimes less to see. This becomes increasingly apparent in this chapter, on Shape, and chapter 8, on Effort, where we endeavor to understand *relationship* and *motivation*, respectively. Moreover, the perception of these ideas is more subjective, more dependent on the experience of the viewer. In training movement analysts, a huge emphasis is placed on learning one's own bias and preferences—in part to recognize, acknowledge, and try to challenge our bias and in part to understand more deeply that our own experience, mood, and prior training changes how we observe the world.

Studying motor development has helped shape the BESST System: seeing simpler patterns in growing children helped name and codify the descriptions and principles that make up the system now. In our own work, we likewise find that robots afford a similar opportunity—one that has helped us refine and reframe some aspects of the work. In developing technology with these ideas, we have reached a clearer understanding of what the elements of movement studies are, can, and should be. This will become apparent in this chapter, for example, where we suggest new ways of discussing "shape quality" based on observations of successful robots that are teaching us new things about expression through movement.

7.1 Foundational Arrangements of the Body in Space
Beginning to explore the shape of the body

Shape is concerned with perceiving and experiencing the body as it changes form and arranges itself in space. As such, Shape is an outgrowth of the relationship between the Body and Space components, defining the bridge between the mover and its environment. Changing form can be experienced and perceived in a *self-to-self relationship* (in both the content and container of the body) and in a *self-to-other relationship* (objects and agents in the environment, as well as the environment itself). Shape can be experienced and perceived in static patterns, or so-called **still shape forms,** as well as in dynamic patterns, or so-called **primary patterns of shape change.**

The BESST System identifies five primary still shape forms: *pin, ball, wall, tetrahedron,* and *screw.* These forms are described in box 7.1 and symbols for each are provided in figure 7.1. Although the living body is never "still," we perceive forms in it; likewise, when a machine sits in the "off" state, the form of the body already takes on meaning. Either through one of the canonical forms listed in box 7.1 or some other shape, the still shape may imply a sense of mobility or stability, jagged or smooth edges, ample or lean dimensions, and so on. For example, the wall and tetrahedron are often associated with stability, while ball and screw are often associated with action or mobility.[1] People often exhibit still shape forms that are particular to their personal expression or movement signature, creating an association with one of these familiar forms, which affects our perception of that person and how they carry themselves.

There are two kinds of primary patterns of shape change in the Shape component of the BESST System: the body can have a *convex/concave* relationship and a *gathering/scattering*[2] relationship (see figure 7.2). The dynamic pattern

---

Box 7.1
Still Shape Forms

- **Pin:** An elongated and narrow shape (e.g., a tall, thin villain like Jafar in Disney's *Aladdin*)
- **Ball:** A rounded, spherical or circular shape (e.g., the typical rendering of Santa Claus)
- **Wall:** A flat, wide shape (e.g., a drill sergeant standing tall and broad)
- **Tetrahedron:** A triangular or pyramidal shape that has a wide base of support (often four points of contact) narrowing to a point (e.g., a person in a seated meditation)
- **Screw:** A twisted form, serpentine in nature (e.g., a Greek sculpture of an athlete preparing to throw a discus)

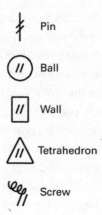

Figure 7.1
Symbols for still shape forms. Each symbol uses the Shape component symbol (a double slash) overlayed with an abstract representation of the form.

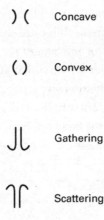

Figure 7.2
Symbols for primary shape patterns.

of concave/convex is seen in the motion of the body—especially focused on the core—opening or closing itself to the environment. Gathering/scattering identifies the relationship of the mover—especially focused on the limbs—acting toward or away from the environment (or elements within it). Our sense of self as distinct from the environment—although the mover is an entity in its own environment—supports the expression of these dynamic patterns of shape change.

These relationships are framed as pairs because if one is present, so is the other. When a concave shape is found (e.g., on the posterior surface in the lumbar curve of the spine), a convex shape accompanies it (e.g., on the anterior surface where the lower vertebrae cut into the internal viscera of the abdomen). Likewise, when gathering, with the inner surface of the palms curling in toward the mover, the body is also scattering on the back of the hands, which are opening away from the environment. Both of these foundational patterns of shape can be further differentiated into specific shape expressions and types of form change, which will be discussed in sections 7.2 and 7.3.

Examples of the convex/concave primary pattern of shape change include stretching in the morning to offer the front surface to the light (where convex is foregrounded) and bending over to reach your shoelaces (where concave is foregrounded). Examples of gathering/scattering include collecting a pile of laundry into one's arms (where gathering is foregrounded) and spreading grass seed (where scattering is foregrounded). Both types of change reveal dynamic, active relating to the environment for different expressive and functional tasks.

Embodied Exercises

- **Still shape forms:** In this exercise, you will explore still shape forms through common images.
  - Return to the images used in box 7.1 and listed here to "try them on" by imitating the still shape form they suggest. Notice your experience with, associations with, and relationship to the environment in each one. Does any one of them feel more familiar or comfortable?
    - *Pin*: A tall, thin villain like Jafar
    - *Ball*: A rotund, jolly Santa Claus
    - *Wall*: A drill sergeant standing at attention with legs spread and hands on hips, elbows to the side
    - *Tetrahedron*: A person in seated meditation, ankles crossed with knees bent and wrists resting on the knees
    - *Screw*: A Greek sculpture of an athlete preparing to throw a discus
  - Can you find examples of each of the still shape forms that create different associations than the ones here—while still revealing the form

itself? Do the associations you experience in the following change your sense of familiarity or comfort with the shape form?

- *Pin*: Alberto Giacometti's famous sculptures of "elongated people"
- *Ball*: Orson Welles portraying the bad cop Hank Quinlan in the movie *Touch of Evil*
- *Wall*: Cheerleader holding pompoms with hands on hips and legs spread wide
- *Tetrahedron*: A Graham technique trained dancer in the seated "fourth" position on the floor
- *Screw*: A runway model walking away from the camera, twisting her upper body back toward the camera's viewpoint, as she throws a wink over her shoulder

- **Relationships to the environment:** Self-self, self-other (self-one, self-many)
  - As you are reading this, notice what small movements you are making—like tapping your foot, biting your lip, twisting your hair. What adjustments are you making to remain comfortable? When do you find you need to make a change, and why?
    - This series of prompts should help you identify a self-to-self relationship, where you are creating changes in the shape of your body and its relationship to itself.
  - Now, having noticed that, pay attention to the relationship of your body with the object you are in a relationship with (a chair, for example). Make a conscious change in your form in relationship with that object. What helps you to bridge your form and self to that object?
    - This should help you identify a self-to-other relationship, where you are creating changes in the shape of your body and its relationship to a particular object in the environment.
  - If you were to convey your experience to a room full of people, how would that change your form and support your connection to that large group?
    - This should help you identify a self-to-other relationship, where you are creating changes in the shape of your body and its relationship to multiple objects (other humans, in this case) in the environment.

- **Primary Shape Patterns:** This exercise will help you explore two modes of primary shape patterns.

  o *Concave/convex:* From a neutral standing position, drop your chin to your chest and then allow your head to move down toward your feet. Pay attention to the front surface of your body. How is the form changing as you take this action? What about the back surface? Notice that the front of your body is becoming move concave, but the back surface of your body is becoming move convex. Now reverse the process. Allow your tail to drop toward your heels and roll back up to standing. Notice the back surface becoming more concave and the front surface becoming more convex. How do these interrelated changes in your form change your relationship to the environment?

  o *Gathering/scattering:* From wherever you are reading this right now, decide that you are cold and tired. Gather a blanket and pillow from your environment, bring them toward you, and arrange the objects so the blanket is wrapped around you and you can rest your head on the pillow. Notice the changes in your form as you gathered those objects toward you to use for warmth and comfort. Now decide that you are hot and awake; scatter those same objects away from you back into the environment. Notice the change in your form as you put those objects away from you.

7.2 Modes of Shape Change
Introducing ways of connecting to the environment

In establishing our palette for Shape, **modes of shape change** illuminate specific and distinct ways that the body's form changes in relationship to the self and the environment. Each mode reveals a progression in differentiation of awareness toward the body/space relationship. First, **shape flow** is a self-referential pattern of shape change and is often seen in self-soothing, unconscious changes in form, and shifting and reorganizing the body; the relationship is between self and self. Movements of shape flow often occur in moments of preparation and recuperation. By contrast, **directional shape change** is a moment of connecting to—or bridging to—the environment. Finally, **shaping** is about accommodation, adaptation, and a more complex interaction with the environment. These terms are outlined in more detail in box 7.2 and figure 7.3.

Box 7.2
Modes of Shape Change

- **Shape flow:** The mover's form changing relative to the self. Often described as "me to me," it can be seen as "checking in with the self" and "readjusting for comfort." Examples include hair twisting while reading, stretching and rubbing out body parts to recuperate, a breathing check-in with the self prior to executing an event. The locus of control is "me," where "me" is often undifferentiated from the environment. This mode is supported by and closely related to the foundational body ideas of flow-sensing and weight-sensing.[3]

  - Shape flow is connected to change of the body, especially as related to the breath, through the following terms:

    - *Growing/shrinking*: Undifferentiated shape change affined with inhaling and exhaling, respectively.

- **Directional shape change:** These are characterized as bridging forms that occur in an arc-like path or a spoke-like path. This is described as "me to you" or "me to the environment." While the locus of control is still "me," there is a differentiation of what is "me" and what is "not me" (the environment). Thus, this mode is referred to as recognizing "you." These kinds of shape changes are often communicative and clarifying in producing outcomes (imagine giving someone directions, ringing a doorbell, or swiping a cell phone). This mode is supported by the body actions of flexion/extension and adduction/abduction.[4]

  - *Arc-like*: This mode of bridging to the environment often describes an arc and an edge or boundary. Using the image of a bicycle wheel, this mode of bridging would be moving along the rim of the wheel with a body part, especially one at the distal end of the body. Often, this may be seen in examples like directing traffic or swiping on the surface of a touchscreen.

  - *Spoke-like*: This mode of bridging to the environment often describes a more linear shape, coming straight from the mover to the environment. Using the image of a bicycle wheel, this mode of bridging would be moving a body part along the spokes of the wheel, connecting from the center to the rim. Often, this may be seen in examples like ringing a doorbell or tapping on a touchscreen.[5]

- **Shaping:** This mode describes changes in the form that accommodate and/or adapt and, as such, is often referred to as being about "us." There is a sense of engagement and cooperation with an "other." Examples of shaping

(*continued*)

Box 7.2 (continued)

include engaging with tools and objects (molding a piece of clay into a pot, twisting the lid off a jar) or with others (embracing a friend and changing your shape to accommodate his shape and form). This mode is often supported by the basic body action of rotation and can occur in the limbs and/or core.[6]

- Shaping is connected to the deformation of the body's innersphere, called **inner shaping,** employed most centrally during breathing and differentiated in each of the three ordinal dimensions as follows:

  - *Lengthening/shortening*: Torso shape change in the vertical dimension
  - *Bulging/hollowing*: Torso shape change in the sagittal dimension
  - *Widening/narrowing*: Torso shape change in the horizontal dimension

The modes are seen to increase in complexity and animacy and have even been likened to patterns of progression in human motor development. For example, the act of shaping requires a more developed motor skill—as well as mental capacity—than that of shape flow. Like all aspects of movement we have described so far, the modes of shape change are perceived and experienced in context, and the meaning will change based upon where, what, and when we perceive their expression (or function). Possessing a way to describe different kinds of relationships with the environment gives us more resolution in identifying evidence to support a particular conclusion about the meaning of a particular movement phrase.

Figure 7.3
Symbols for modes of shape change. Each symbol leverages the Shape component symbol (a double slash), and the symbol for directional mode of shape change is a superposition of the symbols for spoke-like and arc-like. Shape flow also relates to the flow axis on the Effort graph (discussed in section 8.2 in chapter 8).

Embodied Exercises

- **In the kitchen:** Explore each of the modes of shape change in the context of a kitchen as given in this exercise. Each provided example suggests an action that may be best suited (according to our experience) to help you find these patterns in your body, but you may successfully find each pattern in each example—these are rich, complex tasks!

  - Shape flow
    - Attend to yourself as you prepare to cook: adjust your apron, settle your glasses, and so on.

  - Directional
    - *Arc-like*
      - Stir a pot of soup.
      - Feed yourself.
    - *Spoke-like*
      - Spear a piece of food with a fork.
      - Take a hot tray out of the oven.

  - Shaping
    - Take a pot and wipe it out.
    - Open the lid of a jar.

- **In conversation:** Explore each of the modes of shape change in an imaginary conversation (this can also be done with a partner).

  - Sit in front of a mirror and have a conversation with yourself using only "Yes," "No," and "Maybe."

  - Try each mode (shape flow, directional, and shaping) with each word in turn. For example, see how shape flow makes your "Yes" response difficult to hear or believe; see how a spoke-like directional mode of shape change makes an especially forceful "No"; see how shaping makes "Maybe" feel especially persuasive.

7.3   Shape Qualities
Complex relationship of complex bodies to complex environments

Section 7.2, on modes of shape change, describes distinctions in how a mover relates, or bridges, to the environment. This section explores a deeper idea of relating innersphere to kinesphere (or "inner space" to "outer space"), expressing intent, and sharing (or expressing) oneself with the world. We understand **shape qualities** as the most salient patterns in the relationship of body to space, relating complex changes in the container of the body to the environment and revealing a heightened relationship between movers and their environments. Analogously, the next chapter, which will introduce "effort qualities," addresses the relationship of the body to time.

Shape qualities are highly differentiated patterns of shape that leverage our three-dimensional bodily form. The BESST System identifies shape qualities as being related to three-dimensional space, but rather than being about space in simple terms (e.g., "up" can be revealed by myriad different body parts in motion), shape qualities utilize a full activation of the body to reveal a changing form in the environment. The experience and perception of shape qualities recognize our inner space and inner shape mobilizing in an externally observable shape pattern. Shape quality is a heightened moment of salience between a mover's body and its environment, which in humans is especially noticeable in the movement of our spine, muscular core, and breath, but it also can be seen to some degree in distal and proximal interactions, especially in less expressive platforms (like the robots discussed in section 7.4).

The list of shape qualities (outlined in box 7.3 and figure 7.4) are **rising, sinking, spreading, enclosing, advancing,** and **retreating.** Each of these involves complex, three-dimensional bodily motion (in humans, this is most saliently viewed in changes in arrangement of the core) that aligns with a dimensional idea in space.

To consider shape qualities in context, imagine a ballerina performing Juliet seeing the dancer playing Romeo onstage. She walks toward him, but she also extends her spine upward and forward, letting her breath expand into the space. She is rising, advancing, and even spreading, inhabiting the most indulgent state of shape quality in order to express, in this context, her joy at seeing her soulmate. In contrast, upon seeing Lord Capulet, the story's villain, Juliet may also walk toward him, but she would do so with recoil, condensing her core away from this character, exhibiting the shape

> **Box 7.3**
> Shape Qualities
>
> - **Rising/sinking:** Up/down in the vertical dimension[7]
> - **Spreading/enclosing:** Expanding/contracting in the horizontal dimension, e.g., expanding (or contracting) away from (or toward and across) the midline with one or both sides is spreading (or enclosing)
> - **Advancing/retreating:** Forward/backward in the sagittal dimension

quality of retreating and sinking in order to reveal, in this context, frustration, dislike, and even hatred for this character.

Dancers practice manufacturing such expression; it may well be that these dancers do not get along at all in rehearsal, but onstage, they use their facile bodies to create salient movements that tell a story. They engage their inner complexities as fully as they know how in order to bring authenticity, artistry, and allure to onstage moments. In real life, this may not happen with the same salience—or it may happen in ways that feel less manufactured and more authentic because real emotions are involved. Shape qualities help reveal the continuum between inner and outer space in human movers. We often forget that our organs and their fleshy, muscular containers, supported by our facile spines, experience movement too, and shape quality is this idea that connects the shape and shape change of our core to externally visible expression.

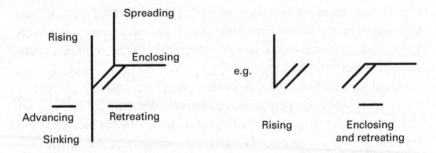

Figure 7.4
Creating symbols for shape qualities. Drawing subsets of the graph shown at left produces symbols for individual qualities and combinations of qualities (e.g., "rising" and "enclosing and retreating," shown at right). See appendix A for a complete list of combinations.

Embodied Exercises

- **Innersphere, breath, and shape quality:** In this exercise, you will find core support for shape qualities through breath.

  ○ As you inhale, feel levity through your abdomen, and as you exhale, feel grounding through the abdomen. This aligns rising with inhaling and sinking with exhaling. Try the reverse to see which feels more natural for you. Take note: the diaphragm moves up when you exhale, relieving your viscera of some pressure, and down when you inhale, compressing your viscera to make room for air in the abdomen.

  ○ Try associating your inhale with spreading and your exhale with enclosing (and vice versa).

  ○ Try associating your inhale with advancing and your exhale with retreating (and vice versa).

- **Hug a teddy bear:** In this exercise, you will interact with a favorite stuffed animal to experience how changing your form changes the experience of the interaction.

  ○ Hug a favorite stuffed animal with each shape quality listed here, and try creating the symbol from the graph in figure 7.4.

    - Rising and advancing
    - Retreating
    - Spreading, retreating, and sinking
    - Advancing and enclosing

Notice how your sense of relationship with the stuffed animal changes as you employ different shape qualities. What works best and feels most satisfying in engaging with the stuffed animal?

7.4 Application to Machines: Expressivity in Natural and Artificial Systems
Counting postures as an approximate measurement for expressivity

As discussed in chapter 1, the ways that machines outperform humans are well characterized. Robots exhibit greater precision and repeatability in their motion than human bodies do in theirs, and they also can exhibit a greater range of forces, torques, velocities, and displacements. But machines—especially automatic or autonomous machines—that match the variety of motion that humans exhibit have not yet been developed. Moreover, they lack many of the moving parts that humans use in our movement profiles; this alone could restrict their ability to communicate through movement.

We have up to this point told promising stories about how movement studies can help create more expressive machines. As we progress into elements of the BESST System that deal with higher-order, more full-fledged expression of complex ideas, it is worth stopping to ask: "Can machines exhibit the same ideas as humans in artificially generated movement profiles?"

For example, take the concept of shape quality introduced in this chapter. In humans, a concept like rising shape quality is typically seen most prominently when there is an articulation of our spines that creates a deep exchange between an idea in our innersphere and our kinesphere, a sharing from a rich inner to a rich outer space of movement. Can robots, which rarely have *any* articulation in their cores and often have relatively simple internal models of the world, accomplish such an act?

To try to answer this, consider the Boston Dynamics BigDog robotic platform. This hulking, quadruped robot has a physical form that does not even quite accomplish the list of features discussed in boxes 4.1 and 4.2 in chapter 4 (e.g., a shape that clearly establishes front/back, right/left, and top/bottom), as it lacks differentiation front-to-back. However, when the object is in motion, it establishes this sense of heading through a pronounced and clear line of locomotion. The device was designed to carry heavy loads for soldiers through varied terrain, and even a decade after its introduction, it remains—along with its suite of follow-up devices, including WildCat and Spot—one of the most successful robots at navigating distinct terrains and environments.

The robot has four legs, each with four active degrees and one passive degree of freedom where the leg can mobilize. The first point of mobility is where the leg connects to the platform, forming a proximal joint. Roughly a third of the way down the leg is another place where mobility can occur, forming a mid-limb joint, and finally two-thirds of the way down is another, more distal point of mid-limb mobility. Each distal end utilizes a round form factor that accommodates many points of contact in diverse terrains.

The four legs work in coordination, utilizing both homolateral (body half) and contralateral (cross-lateral) patterning to mobilize the robot in many directions, but there is no way to actually discern which is the front or the back, as the implement for carrying the load is simply a big, rectangular holding platform. However, some sense of directionality (which is more complex than simply identifying the object's front from its back) is evident through the device's motion as it moves toward, away from, around, and in other ways relative to objects in the environment. That is, we perceive spatial intent when the robot begins to mobilize because of our human experience of spatial intent in our own mobilization. The robot seems to be going somewhere (regardless of whether it is or not) and executing complex locomotor pathways through space as it navigates terrain—similar to what human bodies do, albeit through drastically different means and methods. This sense is heightened by the relatively short duration of the robot's steps alongside human counterparts that seem to step slowly; the robot's pacing is short and choppy next to its bipedal counterparts.

These initial observations concern the Body, Space, and Time components. All these observations are rooted in and related to the human form; seeing the body parts of the machine move through space and time suggests initial ideas about locomotion, direction, and pacing. As we view the robot in a wider variety of circumstances, however, we may see more complex movement ideas emerge.

In one of the most viewed videos[8] of this robot, we see it traverse a wooded hill, navigating chaotic layers of leaves, branches, and trees. This feat alone is a triumph in autonomous navigation, as it is much harder to navigate in this kind of chaotic environment than in the relatively well maintained and homogeneous roadways where cars navigate. But then we see the robot crossing a paved asphalt parking lot partially covered in ice, where a tall male wearing thick, heavy boots kicks the device with great force, causing it to stumble on the slippery ice. Amazingly, the robot maintains its stability, managing not to fall through a series of manic yet intentional swipes of its lower extremities, which twist viciously at their proximal connections to the static center of the device (where the payload would be carried).

For us, emphasizing the heightened environmental factors, we do see a sense of spreading shape quality as the "legs" rip outward to recenter and reorient the device safely upright. Perhaps it is not the same kind of supple, luscious spreading exhibited by the ballerina described in the previous section, but it is a moment of true environmental accommodation, which belies a sense of expression—*if only because we ourselves have been pushed and we ourselves have slipped on ice*; and, as a result, we contain the bodily memory of those experiences, which required complex changes in our own core.

Thus, we can say that in certain settings where the movement of a robot is tightly coupled to its environment, we can see that shape qualities—as well as other ideas about movement—are not limited to the way that humans express them. When we see an object in a situation familiar to us navigate that situation successfully—possibly using different physical means than us—we see, experience, and perceive the same movement idea. Thus, robots may exhibit some sense of shape quality through only their extremities (with their, to date, relatively fixed, relatively inflexible cores), even though humans almost always do so with the help of our mobile, flexible cores.

In fact, because movement studies are so centered around human bodily expression, it is typically taught that shape qualities always manifest as changes in the core. However, in our work with robots, we have found that sometimes we see brief, subtle ideas described in the Shape component in particularly successful or heightened navigation of the environment by a mechanical device (which rarely has a deformable core). Thus, this observation about BigDog is an example of how working with robots has changed our *own* understanding of movement. Moreover, it shows how easily we project our own experience of movement onto other animated objects in our environment (whether they are living, human, or neither).

We can look to the physical complexity of a given body as a way to understand its potential for expression. Prior work informed by movement studies has found trends in this dimension both from an analytic perspective (LaViers, 2019a) and from a perceptual perspective (Cuan et al., 2019a). Figure 7.5 shows how the ideas of shape presented in this chapter can help quantify robot capacity for artificial embodiment. Namely, when viewed as information sources (discussed further in section 9.4 in chapter 9), the multiplicity of possible action is ultimately the driver of capacity for choreography. One way to quantify this possibility is by measuring the number of distinct shapes that a body can form. Such a measure may be key to determining the capacity for a particular platform's artificial embodiment.

Embodied Exercises

- **Exploring animated, onstage characters:** In this exercise, you will consider the expressivity of animated characters. Although these characters are less complex than real animals, they are not constrained by the

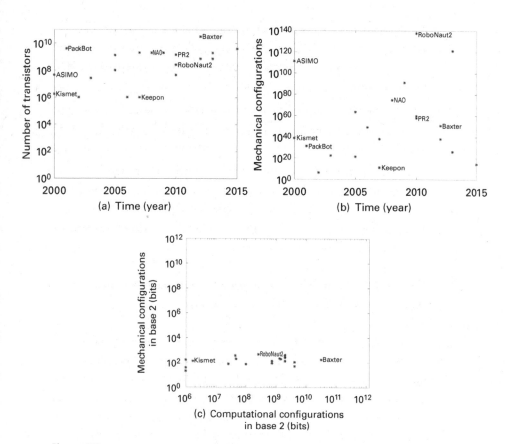

Figure 7.5
Measuring internal versus external capacity for choreography. Internal capacity (a): number of internal transistors on the central processing unit of robots versus their release date; in other words, the number of "shapes" on the inside of these machines. External capacity (b): number of shapes (instantaneous postural configurations) that each robot can make with external sensing and actuation capabilities; in other words, the number of shapes these machines can express in the environment. Comparing internal versus external capacities (c): internal capacity (number of transistors) versus external capacity (in bits) on a log-log plot, showing stagnation in external configurations (or shapes) over the fifteen-year period. This suggests an explosion of internal capacity (as predicted by Moore's Law) versus almost no change in external capacity. From LaViers (2019a, pp. 9–11).

dynamics of hardware, as robots (and humans) are. Moreover, talented artists create their motion profiles *in context* (using environment and narrative to construct situations for characters) for the express purpose of telling a beautiful, expressive story.

- ◦ Watch a movie (or clips from a movie) like the animated film *Toy Story*.
- ◦ Observe which of the animated bodies feels most convincing to you. Why?
- ◦ Are there elements of the convincing characters in which their form changes in a way that seems "successful" in relationship to the environment? Can you identify aspects of Shape that support that observation?
- ◦ What is something that this body *cannot* express? What is something that your *own* body cannot express?

7.5   Exploring the Themes: Shape through the Lens of Self/Other
Understanding the relationship between mover and environment

The Shape component can be seen as working to articulate the bridge between bodies and their movement in space, and, thus, it is useful when illuminating the theme of Self/Other. This duality is a lens through which it is possible to recognize the body as *self* and everything that is not the body (one's own body, that is) as *other*. This "other" could be the environment, objects in the environment, and other moving bodies in the environment. We can bridge, accommodate, negotiate, and change our form to support our interactions and communication with all that "is not us" (other). So from this perspective, Shape can be understood as the continual negotiation of form (self) with the ongoing stimuli of the environment (other) and as revealing a very personal relationship with the world, clarifying the difference between *what is me* and *what is not me*.

As in the other themes we have introduced so far, there is in Self/Other an ever-changing sense of foreground/background. One of the embodied exercises of section 7.3 invited you to interact with an object of high personal value: a favorite stuffed animal. In this exercise, you may have explored how your body moves in response to an element in your environment with which you have a strong relationship. Unlike, say, a piece of garbage, your movement around this item may belie its importance to you: perhaps you cradle it with a complex, swirling sense of shaping, or if you see it hanging on the wall, your spine becomes alert, rising and advancing toward it. Any reaction is valid to your personal and bodily patterns. Using the Self/Other theme, identifying what is us, what is ours, and what is not, we can understand our personal point of view (and how our body may reveal aspects of it through motion) more clearly.

Embodied Exercises

- **Viewing art (relating self to other):** We often think about artists as expressing hidden, specific messages that viewers must decode. But, in fact, some of the most interesting art produces a myriad of varied responses, offering human viewers a chance to discuss, compare, and contrast their reactions. Take Leonardo da Vinci's famous *Mona Lisa*. While learning about its history and the artist's techniques and motivation behind the painting can enrich a viewing, the enigmatic expression of the woman herself captivates and confuses, providing fodder for centuries of discourse. In chapter 9, we will revisit the interaction of context, observer, and object in motion.

  - Visit a piece of art in your community. This could be a public sculpture, a painting in a museum, a movie, or another work.
  - Write the context in which you are observing the piece. Is it raining? Is the venue crowded? Are other people looking at the object (or performers), or is it just you? Is there a scent in the air?
  - Write how you are feeling. Are you hungry? Are people bumping into you? Do you feel like part of the group or are you alienated in some way? Does the scent in the air remind you of anything? What are your past experiences that are coming up as you observe the piece? Which senses are you using to observe it?
  - Now describe the work itself. Do not jump to an evaluation of it ("I like it" or "I hate it"); first, take some time to describe the colors in use, the kinds of lines employed, the textural qualities (this could be in a static object or in movement), and the compositional choices.
  - Finally, consider how the sum total of the components in the last three bullet points affects your viewing and assignment of meaning to the piece. Would it mean something different if it were not raining (or if it were)? Would it mean something different if you were not

so hungry? Would it mean something different if blue were used instead of orange? Would it mean the same to any other viewer?

- **Viewing items of personal meaning (relating other to self):** Repeat the exercise with an object of personal significance. This could be a favorite stuffed animal, a favorite work of dance, an award you once received, or another important part of your life. It should be an object that you have owned/known through the course of many experiences.

  - How do the different elements shift in their contribution to the meaning of this object? Likely, your personal role has an outsized effect on the meaning of this item.

  - If you handed it to someone who did not know you or your history with the object, what do you imagine they would perceive?

Chapter Summary

This chapter covers ways in which a mover may relate to the environment. Outlining modes of relationship self-to-self, self-to-one-other, and self-to-many-others, we highlight static and dynamic patterns that reveal these relationships. We also name shape qualities, intentions that arise out of complex bodies moving in space. In humans, this complexity is most saliently revealed through core change, but machines demonstrate a refinement of this idea, showing how a heightened relationship to environment can produce the same sense through distal action alone. We also note that this relationship is perceived through meaning-making, and thus, the ideas in this chapter are especially sensitive to the specifics of the mover, the context, and the viewer. Moreover, there is a heightened impetus (above that described in chapters 4–6) on the information content of the movement phenomenon. Thus, this category defines higher-order relationships of a body moving in space. The next chapter, on Effort, will outline ideas about higher-order relationships of a body moving in time.

8 How Is the Movement Executed? Movement Quality Enables Shading of Motion (Effort)

The Effort component will address *how* a given movement is happening. Answering this question requires a notion of quality. So far, we have seen components that allow a description of an action such as reaching out for a handshake as: The mover's right arm (Body) reaches forward (Space) over the course of 5 seconds (Time) in a spoke-like manner (Shape). This answers the questions posed by the first four components: What is moving? Where is the movement? When is the movement happening? For whom in the environment is the movement? However, we still do not know as much about the inner motivation of the mover as we could. For example, is the arm thrust forward with a sense of haste and urgency? Or does it reveal a more relaxed attitude? Does it emerge with a sense of airiness or a sense of dense strength? The answers to these questions are often apparent in such a gesture—particularly an act of greeting, such as a handshake—but we have not yet provided a language to distinguish between them.

Effort models the varieties of ways that we try to perceive a mover's motivation (and the ways that we try to communicate our motivation to others). We find it is useful, particularly with regard to understanding machine movement, to examine effort as patterns of dynamic change over time. Here, we emphasize the hierarchical model of the BESST System (see the right image in figure 3.4 in chapter 3) where effort is the result of many bodily activities and relationships. Thus, in this chapter, we also introduce the affinities of the BESST System to demonstrate how effort manifests through complex action in the world.

Effort is one of the most tantalizing components of movement studies for researchers, with many hoping that it will help decode (or encode) the reasoning behind a given movement. Effort can be useful in this process, but it does not offer such answers independent of the other components of the BESST System and the movement's context. As with shape quality, these ideas rely on the manifestation of many components of the body and are revealed through context. We may need to watch a mover for a long time to learn how their baseline patterns modify the execution of any particular, single act. Moreover, we may need to understand more about what is going on in the scene around them. For example: are they greeting a long-lost friend with this handshake or someone who just sued them in court? Although the notion of motivation is an important piece of meaning (note that none of the terms introduced in this chapter or in the rest of part II have an inherent meaning associated with them), answering *why* something is happening remains a difficult task of subjective analysis.

8.1 Basic Effort Actions
Eight commonly identified qualities of motion

In dissecting the notion of *motion quality*, it is instructive to start with a list of already familiar terms. The **basic effort actions (BEAs)** are such a list (see box 8.1 and figure 8.1), offering common verbs that fit into the systematic framework that is presented in the rest of this chapter. As you consider each verb in the list, accompanied by one of its definitions from the Merriam-Webster dictionary,[1] notice the various contexts where you might observe these types of actions. This will help paint the notion of how movement style may suggest a different inner state, or motivation, of the mover.

Box 8.1
Basic Effort Actions

- **Float:** Drift on or through—or as if on or through—a fluid.
- **Punch:** Strike with a forward thrust (especially of the fist).
- **Glide:** Move smoothly, continuously, and effortlessly.
- **Slash:** Lash out, cut, or thrash about with—or as if with—an edged blade.
- **Dab:** Strike or touch lightly.
- **Wring:** Squeeze or twist (especially so as to make dry).
- **Flick:** A light, sharp, and jerky stroke or movement.
- **Press:** Act upon through steady pushing or thrusting force exerted in contact.

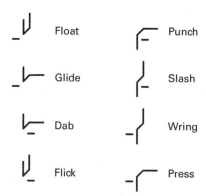

Figure 8.1
Symbols for the BEAs. Each uses the Effort component symbol (a forward slash). These are established from the graph shown in figure 8.2.

The term "effort" is translated from the German word *antrieb*, which also translates as "drive" or "impetus." This word was used to describe an idea of inner motivation—a texture that can be lost in the English term "effort." This list of verbs describes different qualities of movement that can be seen across many activities. For example, we typically associate "wring" with "wringing out a towel" or "wringing one's hands" or even "wringing someone's neck"—but we could also imagine, particularly in contemporary dance, a twisted, wrought solo expressing struggle and strength that did not contain any of those typical activities. That is, "wring" gets at a sense of movement that contains the application of strength, twisting nonlinearity, and continuous, unhurried action.

How does "wring" differ from "flick," "dab," or any of these other listed verbs? We can qualitatively describe the differences (as provided by the definitions) and distinguish contexts where we expect to see one or the other, but the BESST System offers a systematic way to express the differences between these qualities of motion, and many other qualities. This is done through the scheme described in the next section.

Embodied Exercises

- **Trying the BEAs:** In this exercise, you will explore the various BEAs through a kinesthetic channel in functional contexts.

- Look at the list of BEAs given in box 8.1 and make an action with your own body that seems to fit each, focusing on functional tasks that you might encounter day-to-day and gathering any props necessary to fully execute the action. Note the contexts where such an action might typically arise.

  - For example, sliding a shower curtain across a smooth bar might elicit a "glide," or squeezing liquid out of a dish towel might elicit a "wring."

- Looking at the list of contexts created from the previous step, find a different BEA or context that could fit as well.

  - For example, a gliding wave of your hand might elicit a sense of grace and pomp (as used in ceremonies by public figures like a queen). Or, instead of "wringing a dish towel," now try dabbing with the towel, as on a spot of dirt that has accumulated on a shirt.

- Reflect on how these distinct modes of movement give rise to the ability to accommodate different functional and expressive contexts.

- 📹 **Refining expression with BEAs:** The BEAs are also useful for refining and expanding our palette as expressive movers.

  - Develop a phrase that uses "wring," "slash," and "glide."
  - Calling up an arbitrary BEA can be hard to do well. Use a narrative to enhance your expression of these qualities. This can involve a literal story or a series of moods that you inhabit in a sequence that makes sense.
  - How does this idea of expression support your experience moving the prompt? (You can even try recording before and after to witness the change in your movement.)
  - Try repeating this with a few other BEAs.

8.2 Effort (or Motion) Factors

Modeling the quality of movement that belies motivation

The BEAs are described by a combination of **effort factors,**[2] which model how dynamic qualities come together to express some sense of motivation or intent (or the quality with which movers are organizing and expressing their attitude). There are four Effort factors: **weight effort, flow effort, space**

Box 8.2

Effort Factors

- **Weight effort:** Arises out of the fundamental body experience of weight-sensing (which was discussed in chapter 5). It is characterized as expressing intention and sensation and is related to the level of impact we employ in our movement. It relies on an association between one's sense of self and one's sense of one's own mass (and in this way, it is related to weight-sensing). The polarities are as follows:
  - Condensing: *strong* (a forceful way of moving)
  - Indulging: *light* (a delicate way of moving)
    - The expression of weight effort exists as a range between two extremes and can change incrementally without reinitiating action.
- **Flow effort:** Arises out of the fundamental body experience of flow-sensing (which was discussed in chapter 5). It is characterized as expressing progression and feeling. It relies on an association between one's sense of self and one's sense of one's own energy (and in this way, it is related to flow-sensing).[3] The polarities are as follows:
  - Condensing: *bound* (controlling and holding progression)
  - Indulging: *free* (releasing and ongoing progression)
    - The expression of flow effort exists as a range between two extremes and can change incrementally without reinitiating action.
- **Space effort:** Arises through our senses and the bodily ways that we focus on the environment. It is characterized as expressing attention and thought. The polarities are as follows:
  - Condensing: *direct* (channeled singular focus)[4]
  - Indulging: *indirect* (broad scanning multifocal)[5]
    - The expression of space effort must be reinitiated to reveal change; therefore, while a range of intensities might be possible to exhibit, a movement cannot become more direct (or indirect) without becoming something else first.
- **Time effort:** Arises from our relationship to chronological time and our attitude toward it. It is characterized as expressing commitment and decision. The polarities are as follows:
  - Condensing: *sudden* (urgency and possibly accelerating)[6]
  - Indulging: *sustained* (lingering and possibly decelerating)
    - The expression of time effort must be reinitiated to reveal change; therefore, while a range of intensities might be possible to exhibit, a movement cannot become more sudden (or sustained) without becoming something else first.

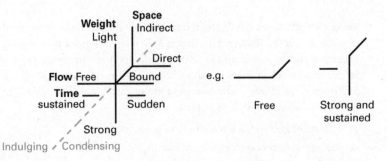

Figure 8.2
The Effort graph. All effort factors and their polarities are shown at left. Drawing subsets of this graph produces symbols for individual qualities and combinations of qualities (e.g., "free" and "strong and sustained," shown at right). See appendix A for a complete list of combinations and their symbols.

effort, and **time effort.** Each factor has two contrasting polarities: **condensing** (making, fighting, going against) and **indulging** (allowing, going with).

To perceive these qualities, we need to notice relative change, so it is when movement changes in energy, tone, or attitude that we perceive effort. Each of the four factors is found rarely, in fleeting moments that punctuate our movement. It is in their combinations, phrasing, and organization that we perceive dynamic expression. The relationship between the factors is organized visually in figure 8.2, which forms the basis of how effort is notated symbolically.

Embodied Exercises

- **Comparing two BEAs:** This exercise will compare two BEAs to reveal what aspect of motion each effort factor aims to describe.

  ○ Dab/flick: Put a damp washcloth on your finger and "dab" at a spot on your shirt or pants, imagining that you have just spilled a bit of food that has been absorbed into the cloth. Then switch, instead imagining that the food is resting on the surface of the cloth and "flick" the food away.

    ▪ How are these experiences different for you?

    ▪ Notice that "dab" and "flick" differ by one factor: the use of space effort. "Dab" has a directed, attentive quality, while "flick" is more indirect and diffuse.

- Wring/slash: Use a damp towel and "wring" out the water; now take the towel and "slash" it in the air, continuing the drying process.
  - How are these experiences different to you?
  - Notice that "wring" and "slash" differ by one factor: the use of time effort. "Wring" uses a luxuriating, indulgent attitude toward time, while "slash" is more hurried and chaotic.
- Comparing dab/flick to wring/slash: What do these pairs have in common, and what is distinct? Notice that "dab" and "flick" have a sense of light weight effort, while "wring" and "slash" use strong weight effort.
  - How are these experiences different for you?
- **Exploring flow:** Use the images given here to experience free and bound flow effort.
  - Imagine you are an old-growth, solid oak tree covered in ice during the winter. A hearty breeze is blowing, causing movement in your branches, which is inhibited by the ice.
    - This image uses a common mechanical failure in trees to conjure an experience of bound flow effort: during icy winter storms, branches can seize up, covered in ice, losing their pliability and breaking under extreme stress.
  - Now imagine you are a young, supple stalk of bamboo in a moist and verdant jungle. A hard breeze is blowing, causing movement that is free and unrestrained. As the wind picks up, you continue to move more and more and more until you are whipping wildly back and forth—but not breaking.
    - This image uses the flexibility and strength of bamboo to encourage you to engage in free flow effort, creating large momentum shifts and an intensely energetic, ongoing movement experience.

8.3 States and Drives
Going beyond the BEAs to the full richness of motion

While the BEAs discussed here are highly crystallized manifestations of three efforts, these occur rarely, punctuating ongoing movement expression. Moments containing three effort factors reveal a heightened intensity and are labeled as **drives.** The BEAs are the articulation of **action drive,** and as such contain no flow effort. However, there are three other drives that do include flow effort, and they are the **transformation drives,** as "action" is transformed into something that includes progression and control. The three transformation drives, which we can identify through the missing effort factor, are **passion drive** (no space effort), **spell drive** (no time effort), and **vision drive** (no weight effort). These are listed in box 8.3, and symbols are provided for each one in figure 8.3. Appendix C further details the individual combinations—also called configurations—within these drives.

As is true with action drive, each transformation drive has eight expressions (varying combinations of the three effort factors contained in the drive), which are often referred to with the same names of the BEAs (e.g., a "passion drive punch," which uses the condensing polarity of flow effort instead of the condensing polarity of space effort). In perceiving and

---

Box 8.3
Drives

- **Action drive:** Combinations of weight effort, space effort, and time effort (having no sense of progression or feeling)
- **Passion drive:** Combinations of weight effort, time effort, and flow effort (having no sense of attention or thinking)[7]
- **Spell drive:** Combinations of weight effort, space effort, and flow effort (having no sense of commitment or decision)
- **Vision drive:** Combinations of space effort, time effort, and flow effort (having no sense of intention or sensing)

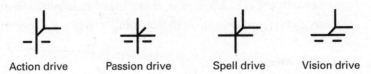

Action drive   Passion drive   Spell drive   Vision drive

Figure 8.3
The four drives in the Effort component. The elements of action drive and all three transformation drives (passion, spell, and vision), illustrated with the Effort graph. See appendix A for the twenty four symbols that correspond to each configuration within the three transformation drives.

experiencing moments of drive expression, the factor that is missing is as important in characterizing the "tone" or mood of the drive as what is present. As with the BEAs, a moment with three effort factors combined into a single expressive event is one of rare, heightened intensity.

Most movement expressions are not quite so loaded: we more often see combinations of only two effort factors. Two effort factor combinations create **states**. There are six of these: **mobile state, stable state, remote state, rhythm state, dream state,** and **awake state,** listed in box 8.4 and with symbols provided for each in figure 8.4. Each state has four possible combinations of the two effort factors contained in the expression. Appendix A offers a complete list of symbols, and appendix C further describes the configurations within these states.

Box 8.4
States

- **Mobile state:** Combinations of flow effort and time effort
- **Stable state:** Combinations of weight effort and space effort
- **Remote state:** Combinations of flow effort and space effort[8]
- **Rhythm state:** Combinations of weight effort and time effort[9,10]
- **Dream state:** Combinations of weight effort and flow effort
- **Awake state:** Combinations of space effort and time effort[11]

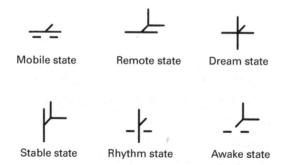

Mobile state   Remote state   Dream state

Stable state   Rhythm state   Awake state

Figure 8.4
The six Effort states. The two factor combinations of the six states, illustrated with the Effort graph. See appendix A for the twenty four symbols that correspond to each configuration in the six states.

It is possible to notice in this list that there are pairs of states that can be seen as contrasting with each other and are named as such (e.g., mobile and stable states). We refer to these pairings as "oppositional states," and one inroad into experiencing each state's distinct qualities is to explore them in these contrasting pairings.

Most movement expressions are a sequence of multiple state expressions that fluctuate and modulate and might become punctuated by drive expressions. As such, they are constantly coalescing in various configurations, and as with all movement events, our experience and perception of effort expression are context dependent. It is also important to note that because effort expression is understood as manifesting an inner attitude or intention in bodily movement, there is also a difference between what is experienced and what is perceived—which are both processes unique to every individual. This means that a mover may be experiencing something different from an observer, especially when those two individuals have significantly different prior experiences, including training, culture, and situational awareness.

Thus, studying movement quality and its performance can improve communication, allowing a mover to better translate intent to others and to better perceive the intentions of others. Sometimes watching the same movement event over and over (e.g., on a recorded video) is required to come to a consensus about what is being expressed. Such refinement in performance and observation is the process of movement analysis—and an inherent element of any rehearsal in the performing arts.

Embodied Exercises

- **Using familiar images to invoke effort:** Each bullet point will provide you with images often (but not always) associated with these states; use them to explore movement qualities. (See appendix C for a complete list of the configurations.)
  - Being startled by something and freezing in response (bound and sudden configuration of mobile state)
  - Kernels of popping popcorn (free and sudden configuration of mobile state)
  - Gasping with a shiver (light and sudden configuration of rhythm state)

- ○ Filled with rapture (light/free/sustained configuration of passion drive; a passion drive float)
- ○ A witch cursing her sworn enemy with an incantation (strong/bound/ direct configuration of spell drive; a spell drive punch)
- ○ Surveying a plot of land (bound/sustained/indirect configuration of vision drive; a vision drive wring)

- **Understanding your personal relationship to effort:** Create your own personal images to help invoke different configurations of effort. Such an effort bank is an essential piece of training movement analysts.

8.4 Affinities between Components
Relationships between components in the BESST System

We are coming to the end of part II, where we have presented each component of the BESST System in turn and, for the most part, in isolation. This works well for the linear arrangement of this book, but it has left out an important concept of the system. In aiming for holistic analysis of the experience of human movement, each component of the system shares **affinities** that highlight how concepts from different components reinforce and support one another. Affinities arise from our body-based movement experience. For example, as we move downward with the pull of gravity, the center of gravity in our lower body sinks and supports, through this bodily movement, our ability to express strong weight effort. We now briefly introduce this advanced concept by discussing Effort's relationship to each of the other components.

8.4.1 Effort-Body
The idea that Effort and Body share affinities has already been evident in some of the entries in the list provided in box 8.1, where certain body parts are mentioned or implied. For example, "punch" is strongly associated with a hand (body part) that is contracted (basic body action) and clenched (muscular tension), forming a fist. While we have discussed that punching can occur in other body parts through any sort of sudden, direct, and strong action, it is clear that actions with a clenched, closed fist read even more clearly as a "punch." Moreover, the physical body-level support offered by contraction is evident not only in "punch" but in other condensing qualities: "wring," "slash," and "press" are all supported through the engagement of body weight via similar muscular activation. Contraction

of large muscle groups—for example, engaging the quads by lowering to the ground, curling the biceps even during overall extension of the hands into space, or contractions in the core—support the feeling that the mover is engaging the full weight of the body and exhibiting strong weight effort.

8.4.2   Effort-Space

Likewise, particular uses of concepts in the Space component can support those in the Effort component.[12] This is in part due to the geometry of our bodies: they have developed in an upright form with bilateral symmetry, creating a somewhat consistent relationship with space. It is also in part due to the shared context our movement occurs in: gravity pulls all moving bodies on Earth *down*. This relationship is formalized in box 8.5 where affinities are represented with a tilde (~), which can be read as "is affined with" (taking inspiration from mathematics where such a symbol might mean "scales with," "is proportional to," or "is congruent to").

The affinities in box 8.5 name relationships where particular spatial pulls enhance the expression of a particular movement quality. For example, when moving around low to the ground, the crouched, contracted position of the body and the required muscular attitude that is assumed (the muscles are really feeling the weight of the overall body here) create a natural affinity for accessing strong weight effort. Thus, spatial pulls in the cube affine with the BEAs as described in box 8.6.

Box 8.5
Affinities between Space and Effort (Dimensional Scale)

- Vertical dimension and weight effort
  - High ~ Light weight effort
  - Low ~ Strong weight effort
- Sagittal dimension and time effort
  - Forward ~ Sustained time effort
  - Back ~ Sudden time effort
- Horizontal dimension and space effort
  - Right (with right side leading) ~ Indirect space effort
  - Left (with right side leading) ~ Direct space effort

Box 8.6

Affinities between Space and Effort (Diagonal Scale)

- Float ~ Right-forward-high (with right side leading)
- Punch ~ Left-back-low (with right side leading)
  - Completing these two actions in sequence also serves to highlight their contrast with one another; each polarity of effort factor reverses as the body crosses abruptly through the central pathway of the diagonal, termed the **float-punch diagonal,** which involves switching from an entirely indulging to an entirely condensing BEA.
- Glide ~ Left-forward-high (with right side leading)
- Slash ~ Right-back-low (with right side leading)
  - This diagonal, the **glide-slash diagonal,** also reverses each effort factor and aligns with the spatial affinity, heightening the expression; here, however, the space effort factor's polarity is switched with respect to the other two factors: "glide," which is indulging in weight and time, has condensing (direct) space effort, and "slash," which is condensing in weight and time, has indulging (indirect) space effort.
- Dab ~ Left-back-high (with right side leading)
- Wring ~ Right-forward-low (with right side leading)
  - In the **dab-wring diagonal,** the right arm crosses the body affining with the directness brought on by crossing and closing the body, opening to heighten the sense of indirect space effort required for the expression of "wring."
- Flick ~ Right-back-high (with right side leading)
- Press ~ Left-forward-low (with right side leading)
  - Finally, the **flick-press diagonal** finishes the scale with a quick, open body that closes with sustainment as it passes down through the vertical dimension, switching from light weight effort to strong weight effort.

8.4.3 Effort-Time

The concept of dynamosphere (introduced in chapter 6) helps to explain affinities between Time and Effort. The Time component allows us to articulate the difference between duration and time effort, which are notably affined. An action of relatively short duration may be more likely to be seen as having sudden time effort, and vice versa. That is, there is an affinity between short and long duration and sudden and sustained time effort, respectively. However, these are not the same aspect of movement—and we can see relatively long durational movements as having a sense of urgency in their motivation, representing an example of a **disaffinity,** where two contrasting movement ideas coexist—and even contrast one another. Rather than heightening the expression of each, this contrast can create a notable expression as well.

Another important relationship to note is that of phrasing and condensing effort qualities. While any of the types of phrases listed in box 6.6 in chapter 6 can be perceived through emphasis, loading, and accent, there is a particular way in which we more easily perceive "accent." For example, impulsive phrasing (where the beginning of the phrase is emphasized) is most easily perceived when the first action in the phrase happens with condensing effort (i.e., a punch is often better at creating contrast than a float). However, a float can feel like emphasis inside the context of several actions of condensing effort. So, while it is not necessary for a short duration, sudden, strong, direct, and/or bound movement to occur in order to see patterns of phrasing, it is especially easy to perceive when an accent in the phrase occurs if these features are present at the same time.

8.4.4 Effort-Shape

As we have established in the hierarchical model of the BESST System, all of the components help to express each other, building up to concepts like shape and effort quality, which are most closely linked to relationship, intent, and meaning. Shape quality, which in adult human movers is typically manifest through changes in the spine, breath, and core musculature, helps express effort quality. One way of seeing this affinity is to consider how one would express light weight in their movement: there is an almost immediate sense of rising that helps communicate this idea. At the same time, consider when we feel a sense of lightness through our experience. Often this occurs in moments of joy or rapture when we feel a delicate, fleeting sense of release, or escape, from gravity. That is, the expression *and* experience of light weight in bilaterally symmetric bipedal bodies that work to stay upright in a gravity field correlate with vertical expansion through the core. This idea is directly embedded in the forms of the Effort and shape qualities graphs (figures 8.2 and 7.4 in chapter 7, respectively), which can be superimposed to reveal these affinities, listed in box 8.7.

Box 8.7

Affinities between Shape and Effort, Indexed by Spatial Dimensions

- ◦ Vertical dimension
    - ▪ Rising ~ light weight effort
    - ▪ Sinking ~ strong weight effort
- ◦ Sagittal dimension
    - ▪ Advancing ~ sustained time effort
    - ▪ Retreating ~ sudden time effort
- ◦ Horizontal dimension
    - ▪ Spreading ~ indirect space effort
    - ▪ Enclosing ~ direct space effort

The organization of the list in box 8.7 suggests further affinities between effort, shape, and space. Notice that in the diagonal scale with the BEAs, given in box 8.6, another element can be added to heighten the expression of each note in the scale: shape quality. By accessing shape quality, particularly through adding deformation of the core to each gesture with the arms and limbs, the mover can further heighten the sense of each action. For actions involving light weight effort, the mover does well to perform rising (although it is not necessary to see this quality). Box 8.8 spells out these affinities between shape qualities and the three-dimensional pulls in the diagonal scale, but the expression of any spatial pull can be heightened with shape qualities (and vice versa): for example, rising and sinking are affined with "place high" and "place low," respectively.[13]

Other affinities exist (and have yet to be discovered) in the system, which are advanced topics that we will not cover here. Likewise, we have not covered movement scales in additional spatial forms (e.g. the dodecahedron) that are being worked out in the movement studies community. Other areas of expansion and evolution of the system include the Time component—our initial treatment given in chapter 6 is sure to evolve in the coming years—and notation, which we will discuss and extend briefly in chapter 10. Both of these loci for new ideas are being pushed by work with machines.

Box 8.8

The Diagonal Scale (Right Side Leading) with the BEAs and Shape Qualities

- **Float-punch diagonal**
  - Float: right-forward-high with spreading, advancing, and rising
  - Punch: left-back-low with enclosing, retreating, and sinking
- **Glide-slash diagonal**
  - Glide: left-forward-high with enclosing, advancing, and rising
  - Slash: right-back-low with spreading, retreating, and sinking
- **Dab-wring diagonal**
  - Dab: left-back-high with enclosing, retreating, and rising
  - Wring: right-forward-low with spreading, advancing, and sinking
- **Flick-press diagonal**
  - Flick: left-back-high with spreading, retreating, and rising
  - Press: right-forward-low with enclosing, advancing, and sinking

---

Embodied Exercises

- **Spatial pulls in the cube (revisited):** This exercise will revisit the spatial pulls in the cube, adding in the affinities—and disaffinities—between Space with Effort and Shape.
  - Move the spatial pulls in the cube (as described in box 8.8).
    - Use affinities in effort and shape to support each expression of space.
    - Try the scale with the disaffinity in effort and shape as you move to each spatial pull.
  - Pick one spatial pull of interest to you. Practice the affinity in shape quality and effort, as well as the disaffinity. Come up with a scenario that is personally meaningful in each. What do you notice about the differences in these personal scenarios for the affinity and disaffinity?
  - Return to the last embodied exercise of section 5.3 in chapter 5, where you may have created your own scale. How does adding in shape and effort to your execution enhance your ability to feel each pull? (Note that here you may be inhabiting only states or single factors, depending on which platonic solid your scale was in.)

- **"Wring," "slash," "glide" (Revisited):** In section 8.1, we asked you to create a phrase with the BEAs "wring," "slash," and "glide," and you explored how narrative helped create a clear expression of those BEAs. Now, try the phrase again, explicitly adding support from shape qualities.

  ○ Wring: sinking/spreading/advancing
  ○ Slash: sinking/spreading/retreating
  ○ Glide: rising/enclosing/advancing

8.5 Application to Machines: Simulating Effort for Advertisement of Internal State

The holy grail of human-robot interaction?

Roboticists and computer vision researchers have long looked to the Effort component to help generate and recognize styles in motion quality. It helps answer the question of how a movement may vary, revealing distinct meaning in context to a particular viewer. Despite many attempts at quantifying the behavior, manufacturing or simulating Effort is an open research problem. Broadly, there is some agreement that the issue involves the temporal pattern of a given motion: changing the evolution of a movement's trajectory seems to change the motion quality (Nakata et al., 1998; Rett & Dias, 2007; Santos et al., 2009; LaViers & Egerstedt, 2012; Kim et al., 2012; Sharma et al., 2013; Knight & Simmons, 2014; Burton et al., 2016; Cui et al., 2019; Bacula & LaViers, 2021). Our presentation of Effort in this book is consistent with this broad idea, but we hope to add nuance to ideas about how humans express effort: it is not only through changing the trajectory of a given body part over time, as most of these prior publications describe. Instead, the affinities between Effort and the other components of the system describe how many aspects of the body work together to create a sense of inner attitude, intent, and motivation. Effort is fleeting and relies on every other component of movement, the body, context, and the observer to manifest.

Amy's doctoral work developed a method for style that is both generative (a library of motion primitives is sequenced via supervisory control and modulated via an optimal control framework) and interpretive (inverse optimizations can be used to classify and segment real observations of human behavior) (LaViers & Egerstedt, 2012, 2014). Simple, "one-armed" automata can be recombined via composition operators with themselves and supervisory automata to produce more complex sequences of primitives (see figure 1.3 in chapter 1). Trajectories are interpolated between these states and modulated by developing a cost function to track (more or less closely) a particular primitive, resulting in variation of trajectories over time (see figure 8.5). This variation is generated by changing weights in an optimal control problem, where each weight is associated with an effort factor: weighting deviations from a nominal, linear interpolation corresponds to space effort; weighting magnitude of overall acceleration (or input energy) corresponds to weight effort; weighting the magnitude in change of state corresponds to time effort; and weighting the match to the final end pose corresponds to flow effort. These associations may be seen as a choreographic technology, enacting a principled treatment of motion quality into an artificial system. However, they are in conflict with a principle introduced in this chapter—that effort drives are rare, fleeting moments—as the work attempts to create motion that is constantly in a drive configuration.

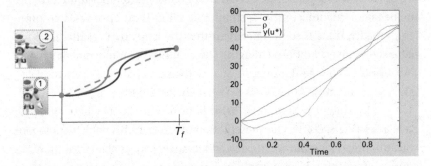

Figure 8.5
Changing movement quality through trajectory modulation. Left: A depiction of two distinct joint angle trajectories starting and ending at the same joint (here, the "shoulder" of a NAO humanoid). These poses may be sequenced as in figure 1.3 in chapter 1. As the joint opens, there are many choices of how the motion may unfold over time; two are illustrated here. Right: Mimicking a real snippet of motion-capture data through inverse optimization (here, one of the degrees of freedom modeling a human leg). The wiggly line (on the bottom at the beginning) is made up of real motion-capture data that has been smoothed; the straight line (on the top at the beginning) is a nominal linear interpolation between the starting and ending pose; the gently arched line (in the middle at the beginning) is a re-creation using inverse optimal control defined by q=1.53, r=0.018, p=0.001 (LaViers & Egerstedt, 2014). In other words, the motion is an artificial embodiment of direct, light, and sudden effort quality.

This work was created in the context of choreographing with and alongside contemporary dancers (LaViers et al., 2014), where it is easy to see how much is lacking in these artificial models of motion style. Namely, motion seems to change its nature on different bodies and in different contexts. A fixed mapping between motion trajectories and high-level, perceptual parameters, while a desirable theoretical idea, falls flat in practice. That is, in developing artificial embodiment for machines, close attention must be paid to the perceptual habits of a wide range of human observers across a wide range of contexts. Some research even suggests the need to consider physiological measures of human interactants, such as recordings of muscle activation (Fdili Alaoui et al., 2017). Ongoing research in this area will surely produce further insights into the complexity and nuance of quantifying the Effort component.

Embodied Exercises

- **Moving with a robot:** This exercise will invite you to find a machine—perhaps you have a robotic vacuum cleaner in your home or can find a video of a machine—and move with it, comparing and contrasting your movement quality through the lens of effort.
  - Find a robot of interest.
  - Imitate its movement quality.
  - Decide which elements of effort are present in the movement of the machine.
  - Based on that categorization, try to move *in the opposite way*. That is, if the machine is bound and direct, try movement that is free and indirect. What kind of contrast does this set up?
  - Use your movement inquiry to refine your estimate of the machine's motion. Try a different hypothesis about the robot's movement quality (e.g., perhaps it is strong and sudden). Try imitating that and then its opposite (e.g., light and sustained) until you are happy with your categorization.

Notice how this exercise is asking you to use your body for research about your own perception and the qualities of machine movement. Effort is one of the most difficult aspects of the BESST System to embody on command in an authentic and externally observable manner. This exercise is not as simple as it sounds and is worth repetition.

8.6   Exploring the Themes: Effort through the Lens of Inner/Outer

Expressing inner life through movement (sometimes)

Effort expression is experiencing and broadcasting an attitude or intention. Our thoughts, feelings, decisions, and sensations—which are often unknown, even to ourselves—are in large part how we know ourselves. These aspects of ourselves are visible only when the inner experience is expressed in outwardly observable movement change, including vocalization of language but also the manner in which we move. As such, the thematic duality of Inner/Outer is one way that we can understand the dynamics of expression.

We have inner experiences of emotion, feeling, attention, focus, thinking, and intuiting that may well be experienced but not necessarily observed (and vice versa). It is only when those attitudes and intentions are expressed in movement change over time that the inner becomes outer or is revealed in a way that is perceivable. It is possible to have a dynamic and rich inner life that is in fact never revealed. In addition, we often participate in forms of subterfuge where we intentionally mislead others as to how we are feeling, what we are thinking, or what our motivations are. It is through effort expression that inner life—or the inner life we want others to think we have—becomes manifest in outward dynamic qualities.

Often, effort has been applied toward the goal of "emotion" generation and recognition, a connection that has been explored but is not yet well understood by psychologists (Tsachor & Shafir, 2019). We hope that the presentation here helps highlight how none of the terms in this system correspond directly to emotion—and aims to emphasize the inaccuracy of such an approach. Assuming that the emotion of a person is visible from the external state *at all* is something that contradicts our own experience: we often work hard to hide negative feelings; many people are described as "hard to read"; and questions like "What are you thinking?" or "Are you okay?" evidence a lack of ability to understand the meaning of someone's movement patterns. Instead, the system aims to outline a coterie of movement options—a mover's palette, so to speak—that can be used, like colors of paint, in myriad ways to create various expressions. The *meaning* of movement expression is individual (on both the part of the observer and the mover) and contextual.

Furthermore, effort need not have anything to do with emotion (consider the lightness that functionally supports placing a delicate teacup high on a shelf overhead). In our application to machines, this is especially true because while machines can change motion styles, they do not have emotions. We do want to endow machines in human-facing environments with variable movement textures to create more useful tools—ones that communicate with their human counterparts either by imitating existing communication styles between humans or employing some novel, device-specific convention—but we do not aim to create life out of inorganic metal parts.

Embodied Exercises

- **Breathing:** Lie, sit, or stand in a comfortable position. Bring your attention to your breath. Focus on experiencing this through the lens of the Inner/Outer theme. Follow your breath as it literally traverses from the inside of your body to the outside and vice versa. Notice how movements happen internally as well externally.

- **Moving with and without an audience:** Revisit the lying, sitting, standing sequence from the embodied exercises in section 6.1 in chapter 6. Go through this sequence several times on your own. Now, put on a camera to record yourself (or bring a friend into the room to observe you) and repeat the exercise. How does it feel to move when you know someone may be watching versus when you are alone? In one situation, you can focus on your inner experience; in another, you may worry about "looking stupid" and modify your experience to present an intelligent or competent persona to the audience.

- **How movement makes you feel:** Try some of the movement profiles listed in figure 8.1. How does enacting these ideas change how you feel or influence your mood? Give this internal investigation a little time to overcome the initial feelings of the awkwardness of the exercise, and notice how your body changes in response to movement.

Chapter Summary

This chapter reviews probably the most famous component of Laban's work, an area that has somehow become one of the most enticing areas of movement studies at the nexus of internal, somatic investigation and external, choreographic presentation: the matter of how inner intent translates into observable bodily changes. As such, many contradictory statements about effort have been made in the literature, and here, we aim to present a practical introduction to the topic that relies on the previously introduced categories: Body, Space, Time, and Shape. Notably, there are specific affinities between these components and Effort that reinforce one another. Moreover, we see Effort as a higher-order relationship that can emerge from a complex body moving in space and time and, crucially, *coping with its environment*. We note that movement quality—or change in movement trajectory—is not enough on its own to create a notion of intent, which relies on a complex cluster of mover, viewer, and context.

III   Translating Movement to Machines

9 Deconstructing Movement: Case Studies in Expression (Answering "Why?")

We want to mold, craft, and design the movement of machines because, as we have seen, movement communicates information. Humans naturally, and almost immediately, attempt to translate information into personal meaning. Part of what training in movement analysis offers is an ability to examine the process of how we come to meaning in order to better see alternative options. In designing robotic movement, we need to distance ourselves from our own immediate, automatic associations between movement and meaning to establish specifications that can be built into machines and broaden our possible solution space.

This chapter presents five case studies to clarify the process of movement analysis, establishing that while machines do not evince all the features of the BESST System, humans can perceive rich meaning from their motion anyway. Each case study is presented as a constellation of symbols from part II alongside qualitative analysis that utilizes the BESST System concepts and principles. In parallel, we will also review more research that has worked to create expressive machines and harmonious human relationships with them—and offer points of view of how movement studies might further dovetail with this work.

The chapter will culminate in modeling the process of communication between humans and machines, drawing a rough parallel to classical information theory. First, an artificial system is modeled as an information source, measured in terms of the number of bits needed to describe it. Then, a human observer is modeled as an information destination that depends on the decoding scheme that it uses to translate a message. Then, context (environmental and situational) is modeled as the channel through which this message is transmitted, which may change the message—for instance, through noise or other disturbances. Finally, we suggest the arts as a testbed for studying this complex process.

9.1 The Duality between Embodiment and Intelligence
The interdependence of movement and thought

The nature of intelligence has been debated for centuries. As discussed in chapter 1, scholars are also grappling with understanding human embodiment. Learning from these scholars, we have used the term "embodiment" to refer to the physical expertise (or "intelligence") displayed by dancers in their studio-based work, by potters in a sculpting course, by singers in a performance of opera, or by football players in their field-based work—to name only a few. All of these "physical" practitioners also benefit from studying strategies and deploying technology to aid their work. This constitutes a concrete way that *embodiment* and *intelligence* are two inherently interrelated aspects of human capability.

In the fields of robotics and artificial intelligence (AI), artificial agents are designed to imitate aspects of human intelligence, although there is a robust and fascinating debate about whether they do: AIs typically excel at things that humans struggle with (e.g., multiplying large numbers and classifying millions of files quickly) and struggle with things that humans excel at (e.g., contextual analysis and making someone feel cared for). In the same vein, sometimes the field of robotics is described as developing "embodied intelligence" (analogous to "artificial intelligence"), but this confounds a key feature of human capability with machine capability and obfuscates the role of embodiment in human behavior. Instead, we characterize work in robotics as developing *artificial embodiment.* Just as the field of artificial intelligence aims to create systems that perform "intelligent" actions, researchers in the field of artificial embodiment (i.e., in the field of robotics) aim to create systems that take "embodied" actions. Such a delineation better describes the inability of machines to reproduce human behavior and sets guidelines for the use of words like "embodiment," "embodied," "intelligence," and "intelligent."

This is not to say that robots do not have interesting movement patterns. In fact, we believe working with robots, along with puppets, kinetic sculptures, animated avatars, and the like, actually sharpens movement studies—just as working with and developing new instruments sharpens music studies. Singers, who create music with their own vocal cords, benefit from this understanding, whether they sing with or play instruments themselves. (We also would do well to remember that all musicians *make music with their own bodies*, even when playing instruments or using computational tools; the division between dancers and musicians is academic.) Likewise, human movers can learn a lot about their own functional strategies and expressive capacities by closely observing the movement of robots and other artificial bodies.

For example, watching the humanoid Boston Dynamics Atlas robot perform a backflip wowed the internet when the video was released in 2017. The viral clip contained the robot jumping on several boxes in series and then rotating itself 180 degrees, parallel to the ground, in order to jump off the last, highest block while rotating 360 degrees in the air, perpendicular to the ground. The device launches itself with stark power before forming a contracted ball that was compared to how human gymnasts tuck to perform complex aerial tricks. In the end, after taking a moment to find stability on both its "feet," the robot lifts both "arms" into the sky in a clear imitation of a gymnast dismount. The move was lauded as Olympics-ready, stunning, and above all, creepily humanlike in various media outlets.

And yet, was it humanlike? How do humans perform backflips? Both acts reduce the moment of inertia of the body, but notice two things that may be initially overlooked but are revealed by movement analysis. Functionally, the robot's flip is not initiated like a human backflip, which is propelled by our hips whipping through space at the end of our flexible spines; Atlas does not have an articulated spine and initiates the movement from its distal "ankle" joints. Expressively, the machine's ending arm raise is notably deadpan and flat, and there is little sense of the jubilance, amazement, and pride that frequently accompany a gymnast's dismount through the carriage of posture (often a performative projection itself, meant to leave a good impression on human judges); Atlas lifts its arms with a stilted, one-note sharpness that is a notable departure from the seeming grace of the in-air, dynamic jumping and rotation.

We see many things that are *not actually there* in a demonstration like this. The idea that it is a humanlike act is in direct contradiction of the single, common feature of all living, lucid humans: breath. Even humans using a mechanical ventilator to live express themselves in the environment through an ongoing deformation of the core, but Atlas does not deploy *any* deformation in the core. Yes, the robot rotates successfully—as does a quarter flipping through the air (though we do not see this as a reasonable approximation of a human backflip)—but what is more intriguing is how it does so in such a different, dissimilar manner to human gymnasts, despite morphological similarities in features described in chapter 4 (see especially box 4.1), such as bilateral symmetry and two "upper" and "lower" limbs.

The moment of raising the arms to finish taps into an expectation that we have about physical performance: we see success and celebration because we have seen it in so many previous scenarios, and we are wowed by the physical triumph of the team of engineers who programmed that action. Likewise, people who own robotic vacuum cleaners often attribute personality and narrative to these remarkably simple devices (which have no such intentional design). Video gamers celebrate the increasing lifelikeness of virtual avatars that, in fact, look very little like real humans—lacking body hair, sweat, breath, and consistent foot traction in their environments, among other differences. How is this perceptual feat accomplished? What are the aspects of movement that need to be present to create this trick? How do we name them? And how do we notice these distinct modes with more regularity and nuance?

This chapter begins to share how the taxonomy introduced in part II can help answer these questions by presenting analyses in a form supported by that taxonomy. Each case study will include a **component constellation,** shown in figure 9.1, highlighting the terms used and displaying each symbol. These constellations are simply ways of gathering symbols together, organized by component, but viewing each symbol shows relationships between the terms (e.g., arc-like and spoke-like directional shape change share all but one feature between their symbols). Chapter 10 explains how these symbols can be used to notate movement using different notational structures to add temporal information and refine the relationships between the concepts and their symbols.

9.2 Expressive Moving Bodies

Movement analysis reveals and enables expression across bodies

Consider a scene from the movie *Star Wars: A New Hope*. A gold, clunky humanoid, C3PO, angles its upper torso toward a young man, Luke Skywalker. Outstretched from the right side of the body is a curled pair of

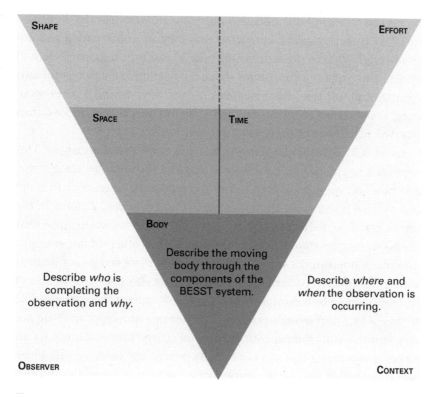

SHAPE

EFFORT

SPACE

TIME

BODY

Describe the moving body through the components of the BESST system.

Describe *who* is completing the observation and *why*.

Describe *where* and *when* the observation is occurring.

OBSERVER

CONTEXT

Figure 9.1

Anatomy of a component constellation. These diagrams offer a template for organizing the concepts inside the BESST System applied to a particular observation or case study. The structure reflects Body as the basis, moving in Space and Time, with higher-order ideas of Shape and Effort emerging from complex interactions of the other components. The areas on both sides of the inverted triangle offer space to acknowledge the role of observer and context in the analysis.

"fingers," with one extended at eye level. The two static "eyes" seem to be looking past the end of this extended "finger." C3PO's human counterpart is gazing similarly and leaning into the device, opening the vulnerable, fleshy part of his abdomen to contact the side of the machine. What is going on in this scene? *Why is the robot moving like that?*

Perhaps you would answer along the lines of "The human and the robot are collaborating," "The pair is trying to figure out where to go next," or "The robot is helping the human." All three of these are reasonable interpretations of the scene. But all of these interpretations (and indeed any others you came up with) are assertions that require evidence to back them up. Why is the typical interpretation that they are collaborating and not competing? What in the scene lends itself to being an important piece of evidence to support a conclusion about the relationship between the two agents? (And, for later: what elements in your own, personal past experience, as well as the environment, make you interpret this evidence in this way? We will discuss these questions in sections 9.3 and 9.4.)

So we ask: what in the scene makes you draw that conclusion? This requires a pause, as you have already assessed the scene in some sense, and now you must go back to before you made that assessment to really *observe* what is going on. This takes practice and expertise. Training in the taxonomy presented in part II is one piece of this, but experts in movement studies further develop the skill to describe what is observed, for example, over many hours in the studio rehearsing a new piece and giving detailed corrections to dancers. Such corrections are most effective—that is, able to be implemented by a dancer—when they are qualitative (commands like "Create a 38.2-degree angle between your forearm and upper arm" are not very effective with human dancers) and objective (dancers cannot fix an aspect of movement that was not actually present and cannot create a new behavior without clear bodily instruction). Likewise, as we will summarize at the end of section 10.4 in chapter 10, movement analysis suggests a similar design process for robots that relies on evidence-based observation and notation, driven by the BESST System's taxonomy, and iterative ideation rooted in understanding the personal and contextual nature of meaning.

In this spirit, let us begin to practice that type of process. Less interpretive descriptions of the scene that we may offer are the following:

- The robot and the human share a common gaze.
- The human is standing close to the robot, craning his neck.
- The robot is *pointing instructively* at something in the distance.
- The robot is *leaning casually* into its human cohort.

These last two descriptions are essential because they identify concrete verbs and adverbs that the robot *seems* to be performing that build up to the larger, higher-level assertions—or *interpretations*—that come more naturally or immediately to a lay observer.

By describing bodily posture, the relationship between the two bodies, and the context of the scene, we are already *performing movement analysis*.

But so far, this analysis is weak. The assertion that the robot character is "pointing" is fully reliant on the contextual and situational features of the scene. Consider how you can confound the "pointing" classification: add a panel of buttons within the reach of C3PO's outstretched finger. Now this static snapshot suggests a different adverb and verb entirely: *carefully pushing* a button.

The taxonomy in part II provides some defense against this issue: it describes the movement rather than classifying or evaluating it. Armed with this type of language, descriptions of the scene can become at once more detailed and less reliant on context. An analysis of this scene, using the BESST System and formatted in a component constellation, is presented in box 9.1.[1]

The case study in box 9.1 demonstrates how the taxonomy enables unpacking of the lay descriptions previously given, which were laden with interpretation, to a description that is less embedded in a particular context or observation. The BESST System facilitates a more granular analysis that organizes itself along particular components with associated principles and uses. For example, the Effort component is used to analyze motivation and intent, while Body helps us understand the physical experience of being a moving body in gravity.

Note that this analysis is not unique or singular: many possible constellations are equally valid. It is simply the process of writing down what seems essential for a particular application. Initially, our analysis included more about the shape of Luke's neck and the relationship of the two bodies side by side (Luke at C3PO's left and C3PO at Luke's right), but upon further discussion, we realized that the meaning of this particular scene in this particular narrative would be the same if these elements shifted. For example, it does not matter if we switch the positions of Luke and C3PO, putting the robot actor into Luke's right side zone instead of his left. The resulting scene could still communicate the idea of collaborating on something that is at a shared point in the distance.

This scene represents many of the aspirations of roboticists today: seamless, harmonious human-robot collaboration. But it is important to remind ourselves that both the robot and the human counterpart are actors reading from a script, supported by directors and other crew members who work hard to make every moment in the movie meaningful to the audience. They are constantly reading the scene, adjusting the set design and reactions of

Box 9.1
Case Study 1: C3PO Helps Luke

- The robot and human share direct space effort, attending to a direction left-forward-middle with their focus, posture, and gesture.
- The human shares an overlapping kinesphere, in mid-reach space, with the robot, engaging in the basic body action of touch, as well as rotation and expansion of the neck, creating emphasis on his focus.
- The robot is bridging to the environment with a spoke-like directional mode of shape change, with all the distal digits of its arm extended in a relaxed gesture.

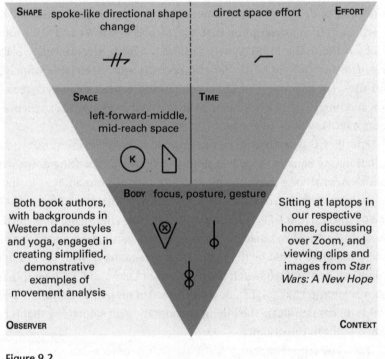

Figure 9.2
Component constellation for Case Study 1.

fellow actors to ensure that a coherent story of relationship is told. It is a galaxy far, far away from reality—it is powered by the performing arts!

The example of two actors casually leaning into one another does not feel like "dancing" or even seem particularly "expressive." It is a rather pedestrian, neutral moment plucked from an action-packed, propulsive plot. But the same taxonomy that describes this moment captures the behavior of social dancing. Using the BESST System to describe casual dancing or bouncing to a rhythmic tune, as in the case study in box 9.2, can capture a distilled essence of this kind of activity in order to explain how simple robotic platforms, such as the squishy yellow BeatBots Keepon or the DreamWorks *Trolls World Tour* dancing Branch doll, achieve a robust imitation of complex activity.

The imitation between bodies described and diagrammed in box 9.2 is a baseline capability in human-robot interaction (HRI): to create social confederates whose changes in posture, gesture, and other expressive elements communicate meaningfully to and from their human counterparts, an ability to imitate the actions that humans make well enough to be perceived as similar is needed. As such, imitation[2] is used as a baseline performance measure in many HRI experiments (e.g., see Gielniak et al., 2011; Simmons & Knight, 2017), and the similar, often unconscious, process of mimicry may be a foundation of human-to-human nonverbal interaction as well (Ashenfelter et al., 2009). Imitation is considered in more depth in chapter 10, which uses notation to clarify imitation across human and machine bodies.

9.3 Expressive Sensing and Perceiving Bodies

We see elements of ourselves in the world

Reading an excerpt from a childhood book aloud, which we can describe as a series of basic body actions of vocalization modified with various Effort configurations, will mean something different to people based on their prior experience (or lack thereof) with the same book. Prior experience will change a person's experience of the information, writ large on the page but digested with nuance by their bodies. A similar process occurs when viewing a piece of art, feeling an unusual texture, tasting an earthy wine, hearing a familiar melody, smelling an exotic perfume, or observing a moving body—a process that is accomplished through a combination of the senses. Sometimes you are drinking wine and sometimes you are not, but

Box 9.2
Case Study 2: Dancing across Bodies

Justin Timberlake (top component constellation)

- The performer Justin Timberlake singing his 2020 collaboration with Anderson .Paak, *Don't Slack*, bouncing to the beat of the song in a social setting (in our particular place and time) utilizes a combination of rotation in the proximal joints (e.g., hips and shoulders), bouncing up and down in space along his axis of length, parallel to his relationship with gravity, which enhances his experience of weight-sensing.
- He will typically try to match the tempo of this repetition with the tempo of the beat of the music, creating a satisfying experience of synchrony of kinespheres with other bodies sharing the space.
- Often he may begin bridging to these other bodies, using arc-like and spoke-like modes of shape change to indicate his awareness of his social counterparts. In addition, changing the dynamic quality—or effort config-uration—of his motion quality can match mood to music and further com-municate socially. In the music video for *Don't Slack,* Timberlake inhabits states associated with passion drive: rhythm, mobile, and dream states.

Keepon (lower left constellation)

- The simple Keepon robot is made of a soft, squishy yellow shell made of two conjoined spheres and affixed to a few linear actuators that telescope up and down in series, pinching the yellow form into various bends that mimic articulation. A popular video of the device shows it moving up and down along the vertical axis matched with the beat of a popular pop song. The subtle folds in the device's soft exterior shell suggest the kind of core deformation associated with shape qualities; we see rising and sink-ing in this action of the Keepon. Rising and sinking are also supported by the movement to high and low, respectively, reflecting affinities between Shape and Space. Although there is no true rotation or spine available to the device, it successfully imitates a human mover and can encourage shared movement in a social setting (Michalowski, Šabanović, & Kozima, 2007; Michalowski, Simmons, & Kozima, 2009).

Branch Troll Doll (lower right constellation)

- Beyond a mess of passively waving purple hair, the Branch troll doll (created as a unit of merchandise for the *Trolls World Tour* movie that featured the song *Don't Slack*) has only one or two actuated degrees of freedom to gener-ate motion: a joint in the center of the doll's torso to which two lower and two upper limbs are affixed. As the motor rotates, it induces a weight shift related to the rotation of a proximally located joint. This action is matched to the beat and meter of *Don't Slack* that plays when the doll is in motion. Although there is no up and down translation in space, the device success-fully imitates a human mover enough to create an entertaining doll.

Box 9.2 (continued)

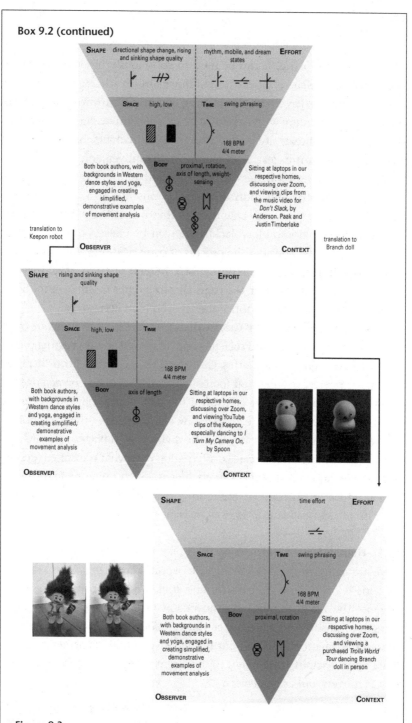

Figure 9.3
Component constellation for Case Study 2. Images of Keepon by BeatBots, LLC/Hideki Kozima/Marek Michalowski, used with permission.

you are *always* moving. It can be hard to consciously observe movement because of its constancy. Notice it right now: authors writing (fingers flying quickly across the keyboard, lips pursing and brows furrowing in concentration, ribs expanding with a dissatisfied exhale) and readers reading (eyes skimming lines of text, sit bones shifting weight in chairs, hands and arms reaching out to turn pages). Both require bodily movement.

Beyond bringing awareness to our constant motion, differences (and inaccuracies) exist in our individual perceptions. In 2015, an image of a striped dress went viral as people realized that two entirely different color profiles (white/gold and blue/black) were seen by viewers of the image of the dress online. Or consider the process of copy editing. Copy editing is a notoriously difficult process, as our sensory systems often see what we want to see, not what is actually on the page, allowing typos to slip by unnoticed. Thus, people do not always notice the environment accurately.

Even more, the conclusions that we draw from the same stimulus are often quite varied. Our cultural and societal context, as well as our personal preferences and experiences, inform this. Consider a close call under consideration by a sports referee. In American football, a "catch" is often characterized as requiring the player to have control of the ball. Sometimes referees make wildly unpopular decisions (e.g., they saw control where others did not, or vice versa); sometimes a play is genuinely divisive, with players, pundits, referees, and announcers in sincere consternation over what transpired (despite extensive replays and multiple camera angles). Such opinions may also be influenced by which team we are rooting for. *How do we know if a football player made a successful catch? How do we know if what we are observing is real?*

9.3.1   Human Senses and Sensing

There are many ways of "knowing" the world, and in movement studies, the body is considered the basis for knowing. We know because we move: we sense the world through our movement (from tiny saccades to larger changes in position to get a better vantage point), and we make sense of sensory information based on our unique, lived bodily experience. Developing our conscious attunement to these senses deepens our capacities not only as movers, but as perceivers:

> The more developed and thorough our capacities for receiving and responding to sensory information, the more choices we have about movement coordinations and body functioning. (Olsen & McHose, 2004, p. 16)

Traditionally, five senses (sight, sound, smell, taste, and touch) are recognized as the primary vehicles by which humans perceive the world; however, the "sixth sense" is that of proprioception, or kinesthetic sensing. Like the other senses, proprioception is another way that we decode and code information—not only through receptors in the skin, but the muscles, tendons, and joints as well. The word comes from the Latin word *proprius*, meaning "own," combined with "receptive"; it is a way that we sense through our own bodies "where" we are in space and in relationship to the environment. Movement is also perceived with sight, sound, and touch (and to a lesser extent smell and taste), all of which are explicitly utilized in somatic practice.

Many robots communicate through more than the visual channel. Researchers in HRI have proposed systems that use sound mirroring workspace tasks for improved localization during collaboration (Cha et al., 2018) and simulating various states of effort (Dahl et al., 2017). A measure comparing systems across several factors, including movement and sound, has also been proposed (Frederiksen & Stoy, 2019). Sound and movement, as well as a soft, cuddly exterior, engaging the haptic channel, were all used in the PARO device, a robotic stuffed animal shaped like a seal that has generated some of the most significant results in experimental treatments with human subjects of any therapeutic or "social" robot (Šabanović et al., 2013; Geva et al., 2020). Temperature is another modality that has been explored, where adding warmth has been shown to increase connection between human and machine (Park & Lee, 2014).

The Theragun is a handheld machine that delivers percussive therapy via a pumping action on the surface of the user's body. It is applied in varying speeds and pressures by the user or a massage therapist. Its purpose is to release muscle tension; help break down scar tissue and "sticky" fascia surrounding the muscles, enhance blood flow and lymphatic circulation, delay the onset of muscle soreness, and improve flexibility after vigorous physical activity. Its design allows it to be easily held and manipulated, and it becomes an extension of the arm/hand of the mover itself, increasing the pressure that is available to be applied to the body greater than the human hand alone could provide.

Through the haptic engagement of the machine with the body surface that it is being used on, the Theragun uses touch to increase kinesthetic awareness. This awareness brings the user into attunement with muscles, bones, and viscera—essentially, the inner contents of the body. This effect mimics foam rolling (where a user presses a part of the body that needs therapy into the sharply rounded surface of a foam cylinder), but with a very different energetic method. The device creates a distinct, strong sense of energy (which can be overwhelming to bring close to the body), which is part of the therapy: users can feel this energy, creating release and a heightened combination of pleasant and unpleasant sensations. Considering the somatic experience of massage with an artificial aid, as done in the case study in box 9.3, which is an example of leveraging a somatic strategy, can aid in designing and testing such a device.

The context of dance, and even the figures in this book, can emphasize a visual absorption of information through movement, but we experience movement through haptic channels as well (as the Theragun example demonstrates). Both the vibration and motion in our own body, as well as bodies in shared space, stimulate our sense of touch. In our own bodies, the skin forms a boundary between the motion of our internal organs and tissues and the motion manifest in the environment. This layer also provides a site for connection to other bodies in the environment. Likewise, all the senses are part of the observation of movement and contribute to our perception of meaning.

9.3.2 Perceptual Habits and Creating Narrative

What happens when a more intelligent agent meets a less intelligent agent? What happens when a more embodied creature meets a less embodied creature? For example, what happens when adults meet a tiny baby? The intelligent, embodied adult makes up a story about what the baby is doing (e.g., "Aww, she's reaching for her mother" or "How sweet, she wants to play with her big sister"). Science offers little evidence that a newborn could have such complex intentions, and yet such a narrative is commonly ascribed to their behavior by playful and loving adults. A tiny reach toward any given individual can have a myriad of possible meanings, most of which fit a desired narrative of familial bliss.

This can mirror what happens when humans meet robots. Regardless of what ideas science fiction (and engineers working in the field) may create about the promise of robotics, human capacity outpaces that of artificial devices today—and probably at least for a very long time into the future—in terms of *both* intelligence and embodiment: our creativity, problem-solving ability, and effective improvisation in the face of dynamic, changing environmental conditions are unmatched and not even particularly well

Box 9.3

Case Study 3: Using a Theragun to Tap into the Haptic Channel

- The theme of Exertion/Recuperation is a way into understanding Theragun's popularity, particularly among athletes, as it enhances the recuperation phase needed after extensive exertion.

- The haptic disturbances created by the Theragun promote bodily healing, shortening the time required to rest and improving performance. In this regard, the machine mimics the basic body action of touch (kneading sore muscles, rubbing areas in pain), bodily behaviors that humans engage in and experience. Operators respond to their own sensations, partly (unconsciously) in a state of shape-flow and partly (consciously) invigorating their musculature, in order to guide the device. In some sense, the device is a mechanical solution for many of the hands-on practices of somatics, which encourage using massage and touch to develop a clearer relationship between mind and body, intention and action. In doing so, the machine allows the user to have increased access to the baseline of weight-sensing and flow-sensing.

- This difference in energetic methods can be analyzed through the Time and Effort components. The vibratory phrasing of the device-user pair creates a rhythmic structure within which the flow of the mover is released in response to the condensing, controlling action of the "hammer" head. In terms of Effort, these energetic changes most often live in configurations of rhythm state: combinations of weight and time effort.

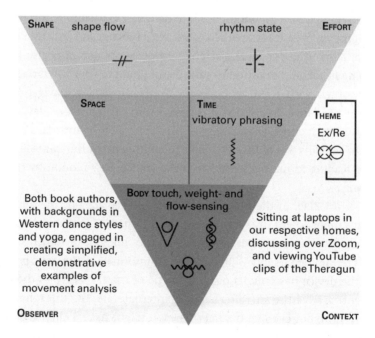

Figure 9.4

Component constellation for Case Study 3.

understood. And yet, when we see machines in the environment, we make up stories just as we do with babies (e.g., "Oh dear, what a pathetic little guy, running into the wall like that" or "How funny, it's waving at me" or "Oh no, it's going to kill us all!"). Each of these narratives may have nothing to do with the design of the device in question but instead reflects the narratives, perceptions, hopes, and fears of the observer. Thus, even robots that are not particularly complex machines may be quite expressive to humans. Or, as the French-Cuban-American essayist, diarist, and author Anaïs Nin (1961, p. 124) wrote: "We see things not as they are, but as we are."

This idea explains some of the impressions that users have of the simple robots that are slowly but surely making their way into consumers' hands. One of the most prolific in-home robot devices is the automated vacuum cleaner, the iRobot Roomba. While other companies have subsequently developed versions of this robot, such as Xiaomi, LG, Husqvarna, Dyson, Samsung, Neato, and Aztech, to name a few, this kind of device is colloquially often still identified as a Roomba, regardless of the developer's name. This commonly used robot, and the user's relationship with it, offer another example where it is possible to illuminate the influence of experience on perception—and how a simple, functional device can be quite expressive.

The Roomba is an autonomous robot designed to vacuum/clean the surfaces it moves over. It is essentially a disc with brushes and a suction capacity that allow it to stir up and remove debris. Some Roombas also have a mopping function that introduces the use of liquid to wipe the surface that it is in contact with. It utilizes lasers and cameras to achieve navigation and obstacle avoidance. In essence, it is designed to clean a floor, alleviating a tedious human chore. Even nearly twenty years after its introduction, the Roomba remains one of the few robots inhabiting many human-facing scenarios, leading to many interpretations of the device's motion by human observers.

What if C3PO, in the example in box 9.1, is replaced by the Roomba? Can this robot "point instructively" or "lean casually"? Probably not. From a mechanical standpoint, it accomplishes the function of moving across indoor terrain remarkably well; but from an information-theoretic point of view, the device has only its heading (angle of travel relative to the environment) to advertise anything about its internal state. Yes, this robot lacks human form, but even with a tail or ears, it might have a chance of indicating changes in internal state through visible movement, but it was not designed for expression.

Forgetting to use the lens of expression when designing a robot has meaningful consequences. For example, soon after deploying mobile grocery store inventory monitor robots, Badger Technologies found that customers preferred its Marty robot model (which has a similar expressive form as the Roomba: it can drive around and change heading, albeit in a larger, narrower form, a pin still shape form) when it had a large pair of plastic "googly eyes" affixed to it (King, 2020). Surprisingly, affixing a cheap, plastic, off-the-shelf decoration to these devices allows a more successful interaction and suggests that a better solution could exist if the lens of expression had been considered by engineers in the design of Marty.

Without the googly eyes, the Marty robot is a large gray monolith that glides in and around grocery store aisles. Due to the placement of its primary collision detection sensor (a mounted scanner), the robot moves with a consistent facing but can rotate with differential drive wheels, allowing it to navigate narrow spaces. However, to the lay human, there is little to differentiate one of its four sides from another: each is flat and gray, differing only slightly in terms of dimensions. Thus, without a clear visual fiducial, it can look like the robot does not keep a consistent direction of travel relative to its body. This ambiguity—or lack of spatial intent—can create an ominous presence for shoppers, who have reported it as "following them around the store" (even though it is not). Thus, the added googly eyes use a familiar visual signal (rather than an unfamiliar red line, say) and, with that simple augmentation, it is easy for shoppers to recognize and identify one side of the machine from another. Now, the robot becomes more expressive to (and more harmonious with) the human interactant as a consistent direction of travel is discernible.

A similar experience occurred for the robotics company Sphero, which made a ball-shaped robot—with perfect radial symmetry—that rolled around nimbly and could be controlled by human operators through a phone app. This initial product struggled to find widespread interest, but when Disney approached the company about making a mock-up of a new *Star Wars* character, BB-8, comprised of two balls, one that rolled on the floor and one that balanced atop this spherical "wheel," the company had a hit device on their hands. As with Marty, Disney's design provided the device with a physical form that identified its body, creating a consistent sense of top-bottom, left-right, and front-back through the asymmetrical decoration that now balanced atop the device. Box 9.4 provides analysis of these three robots: the Roomba, Marty, and BB-8.

Box 9.4
Case Study 4: Mobile Robots in Human Spaces

- From the lens of the Function/Expression theme, the Roomba and Marty provide a service (function) but many users who interface with them ascribe expression to the devices (largely through the way that the devices navigate the environment). This theme highlights the inseparability of function and expression: even devices designed only for functional ends express ideas through their movement.

- For example, when the Roomba is observed in users' homes, the path of travel (locomotion) of the machine can be observed, and its response to obstacles reveals its ability to redirect its pathway to continue on its way. Because it appears to be navigating and coping with the environment, albeit in the most simple and rudimentary of ways, the projection of human action on the part of the observer becomes possible. Moreover, the beeping noises the robot makes (a form of vocalization) when it gets stuck also reveal its inner state—a "cry for help" that allows the human to respond to the machine and further ascribe a narrative or personality to the robot.

- The simple augmentation of the googly eyes to Marty allowed it to take an identifiable form—a face (or a preferred direction for action and sensing) that creates a sense of front, back, right, and left, the spatial zones that the original design lacked. Now, the robot looks like it is *moving sideways to its right*, a much more salient, if odd, movement design that highlights the consistency in the device's operation, helping shoppers feel more at ease with, and a bit entertained by, the machine. The augmentation of the BB-8 with the second ball achieved a similar outcome. The form is now identifiable as having a face, and as such, zones of movement in space that are salient to human viewers.

- To consider the differences between the three robots, look at the still shape forms associated with each: Marty takes the form of a pin, while BB-8 and (to some extent) Roomba are ball still shape forms (Roomba could also be considered tetrahedral because of its stable base).

(*continued*)

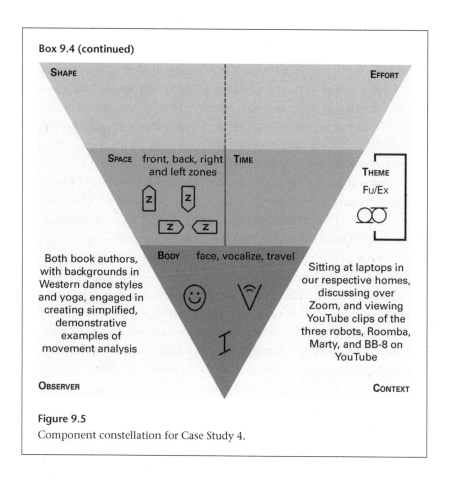

Figure 9.5
Component constellation for Case Study 4.

In addition to the features discussed in box 9.4, Disney's design gave BB-8 a back story and an array of characters (played by actors in the movie) that revealed a personality, narrative, and sense of animacy for the device through their reactions to its scripted behavior. Likewise, we worked with collaborators to affix simple augmentations to the iRobot Create (an open-source, reprogrammable version of a Roomba) to embed the device in a theatrical setting, as shown in figure 6.2 in chapter 6, eliciting reactions from audience members that were made possible by the more perceptible configuration space of the device and the human performers' reactions to it (Pakrasi et al., 2018; Berl et al., 2019). Similar work uses the frame of an animal to inspire mechanical augmentation (Singh & Young, 2012). Bacula and Knight (2019) use both animal inspiration and a theatrical embedding to develop expression with mobile robots. In all three cases, we may say that the new hardware designs employ a choreographic technology and a theatrical embedding to establish a perceptible and consistent orientation in the environment.

Just like the Roomba embedded in a theatrical setting, the scene with C3PO relies on the expertise of artists to make any sense. While it may not be effective at imitating any arbitrary idea, a Roomba, in real life, does not come equipped with a director, cinematographer, scriptwriter, or internal human actor to ensure that the scene reads with clarity. And yet, the Roomba is undeniably expressive to human counterparts (Fink et al., 2012)—even if not *as* expressive as other bodies (per the model presented alongside figure 7.5 in chapter 7). This example demonstrates how it is possible to perceive ideas of "intent," personality, and decision-making that are more complex than the device, which is powered by technology developed in the 1990s, actually employs. This speaks to the notion that wildly distinct bodies different from the human form can be perceived as expressive. The observer perceives the robot making decisions, mapping the space, and negotiating in a way that resonates with the human experience of movement in space. These three robots and the human come into communication through locomotion, navigation, and sound—and, most critically, *the complexity of the human's prior experiences with those modes of movement.*

9.4   Expression in the Environment
How context manipulates meaning

Consider an agent "moving from point A to point B"—a classic task in robotics. Without any additional objectives defined, such as minimum energy, the agent is free to traverse in many, many ways. We may say that the agent *moves freely.* Now, add one complicating factor to the task: the presence of a friend or an observer, which implies a social constraint. To solve this task, the agent should traverse from point A to B in a way that might look reasonable to or even entertain the observer or friend. You might say that the agent now *tries not to look stupid.*

Give the task an environmental context: see two examples in figure 9.6. First, consider that the agent needs to safely traverse a living room environment (right side) in the presence of a friend. Now the task imparts the functional element of avoiding the obstacles in the environment—continuing to have the expressive, social task of looking reasonable to the friend. Turn up the potency of that expressive task, requiring that the agent also communicate "anger" to a friend. Now, a solution might be for the agent to

Figure 9.6
Two distinct environments for movement. At left, a wooded, natural space with thick undergrowth and chaotic obstacles, and at right, a designed, interior space with distinct obstacles. Both inspire different movements and create different contexts for analyzing those movements. Two movers in this space will have kinespheres and dynamospheres joined by the shared context that become part of the environment itself.

stomp with heavy, sure-footed steps and *slash arms wildly in the air*. Such a pattern of behavior would likely alert a friend to a turbulent inner state.

Now return to the original task, getting from point A to point B in an isolated (not social) context in a new environment: a forest, covered in thick overgrowth (left side of figure 9.6). In order to move effectively through this thickly wooded environment, the agent must *stomp with heavy, sure-footed steps* and *slash arms wildly in the air*. Add the observer back to the scene. Does the observer perceive "anger" from the movement pattern? Likely not. Now, the movement takes on a functional lens and likely communicates something like "competency," or maybe "urgency," to the onlooker. The new environment changes the meaning of the motion.

To be complete, note that "the environment" can contain many factors not related to the flora of the biosphere or interior design of the building. People in the environment change what movements mean; the series of events preceding the current event changes what movements mean; and the culture and country in which the movement occurs—which carry their own conventions and politics—change what the movements mean. Thus, we describe all of this as *context*.

Interpreting the meaning of movement as this example has simulated is the goal of a growing area of research in AI. Explicitly modeling context can improve performance (Heimerdinger & LaViers, 2019), but the examples presented in this chapter hopefully highlight how the meaning assigned to any given movement is not fixed. Changes in context (human-facing environments are *dynamic*, the context is always changing), viewer (every human brings a different corpus of experience), and the mover itself (a rich, expressive body) create a moving target for artificial systems.

Yet there is some way that humans are frequently able to come to salience based on the actions of others in the environment, and thus it is a goal of the field of AI to replicate this. We would argue that this is also a problem of artificial embodiment. As we have seen in this book, knowing or interpreting the movements of others is rooted in our own experience. Thus, it is not just a matter of decision-making to determine the state of our counterparts, but actually our physical experience and expression of movement in context inform how we achieve this. That is, we do not observe movements and then come to a decision about what they mean. We experience movements inside complex environments and react to them in order to tease out more information through interaction.

Take, for example, the following movement: *clenched fist to chest, resulting in a concave core shape change around the sternum, and a flexion of the neck, essentially a bowing of the head*. This movement could have multiple meanings in different contexts. For example, it could mean:

- I am so angry that I am pulling my energy in before I explode.
- I am so joyful that I am taking in what I have just received before I open up to the world.
- I am surrendering to subjection and feel powerless to respond.
- I am so tired . . . I just want to hold my heart.
- I have really bad heartburn and am trying to mitigate it.
- I am choking on a piece of chicken.
- I have a stiff neck and upper back today and need to stretch.

Upon seeing the prescribed movement, we may not consciously have any of these possibilities in mind, but our body has experienced many of them, so they are part of our physical experience—our embodiment.

Being aware of this embodiment arises in applications such as using the motion of pedestrians to improve the sensing systems of autonomous vehicles. For example, detecting alarm in a pedestrian could alert the system to an unseen obstacle that is occluded by another car (Afolabi et al., 2018). We will see the same problem in chapter 10: many movement sequences can express the same idea, and many ideas can explain the same movement sequence. This creates technical challenges that so far are intractable. Gathering data to model human movement is one of the biggest challenges in this area. Some researchers have looked to YouTube, television, and the movies for "natural," "in situ," and "unconstrained" movement examples (Luo et al., 2020). Yet, as we discussed in the analysis of Luke Skywalker and C3PO in section 9.2, these samples of movement are highly edited, curated, and selective to develop media that is entertaining, and meaningful, for audiences.

Take, for example, the chore of cooking. Cooking is a daily activity required for human survival; as such, humans have extensive experience of engaging in this movement behavior. The context of cooking and meal preparation often is the kitchen, and most cooking involves imparting flavor through the use of spices to enhance the taste or nutrition of the food that the cook is creating. One famous chef, Emeril Lagasse, developed a particular movement phrase that became an important stylistic feature of his cooking style and helped to bolster his popularity: he would add ingredients with physical flair and punchy ingredient delivery, exclaiming "Bam!" as he added key spices to his food. Part of what makes this movement phrase (a sharp moment of condensing effort quality accompanied by the vocalization "Bam!") so attractive to so many is its play on dynamics and phrasing, which is *unexpected* in the context of cooking.

Lagasse's signature moment involved a pursed, contracted hand directed at the food that he was currently making, a dramatic pause, which included holding his breath, and then an exuberant, even careless, explosion of spice, as well as his breath, as he yells "Bam!" and his fingers flay out into an expanded, released state. The musculature that was holding in his viscera as he pulled in and paused his breath is released along with the spices. This arc is common to many moments designed to create drama: tension building, pausing at a climax, and a quick resolution. Lagasse, who was not a trained mover, expertly created this moment multiple times each show using his own expressive body, as analyzed in the case study in box 9.5.

Lagasse's signature "Bam!" phrase on his cooking shows becomes a salient moment largely through the context in which it occurs. Because he makes this

Box 9.5

Case Study 5: A Profile of Emeril Lagasse

- The "Bam!" moment, which is achieved both with vocalization and breath support, utilizing a held breath and a forceful exhale, generally is employed when adding spices to the dishes being prepared. He dispenses the spices with an impactive phrase that is simultaneously performed through a gesture and a postural shift. The left hand comes up, palm facing the audience, in front of his forehead while his right hand condenses with speed and force to impart the spices into the dish being prepared.

- The Effort component in relationship to the context in which the action is performed is one way of understanding the heightened impact of this movement moment. Through flow effort fluctuations coupled with constellations of awake state (space and time effort), the phrase unfolds and becomes impactive as the heightened moment of the punch (condensed expression of action drive) occurs at the end of the phrase.

- Cooking is often methodical and measured, but Lagasse turns that idea on its head by introducing a movement phrase that surprises the observer. Considering the component constellation here, it is also worth noting what is "not" there: for example, the lack of bridging to the environment using a spoke-like mode of directional shape change, a more typical method of carefully bridging to the environment and adding spices with more decorum that one might expect to see when cooking and using tools. Thus, in designing a cooking robot meant to entertain human observers (e.g., the one currently being developed by Moley Robotics), designers might consider this type of analysis. It is not intuitive, but by using less precise spatial focus—also called "legibility" and "predictability" in some HRI research (Dragan et al., 2013)—a more engaging and "humanlike" robot might be created.

(*continued*)

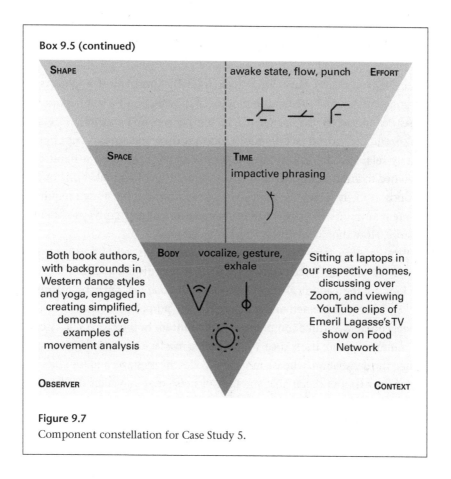

Figure 9.7
Component constellation for Case Study 5.

moment so much about energy, body, and time—but not space—his dynamic quality is different, producing a contrast to his other activities in the kitchen, which require more measured control and menial performance: chopping, wrapping, or tying. To create a moment of entertainment, he essentially "punches" the spices into the dish—with the heightened dynamic moment at the end of a phrase of less punctuated actions. This "surprise" moment becomes salient because it is unexpected in the context in which it occurs.

Lagasse's popularity is not purely a function of the amount of energy or movement he enacted; it has to do with *what* and *how* he communicated. We need a model that allows us to understand that this communication is personal, unique, and contextual. The next section aims to provide such a model to an engineering audience, looking away from traditional motion models and toward traditional communication theory.

9.5   A Model for Expression and Meaning-making

Information theory and the arts

Isaac Newton's laws have long been used by mechanical engineers to
describe, model, and predict the motion of moving bodies. Here, motion is
measured in meters (displacement), meters per second (velocity), or meters
per second per second (acceleration). And yet these measures do not immediately relate to the informative aspect of motion. Imagine a hand outstretched to another human; if that extension of the hand takes one or five
seconds, it likely relays the same *amount* of information (though requiring
different amounts of energy) to a human viewer but may convey a different
message. How should we resolve this as engineers?

Turning away from Newton's laws, because we are, all of a sudden, less
concerned with the forces, torques, velocities, and displacements of a body,
we look to the canonical model of *communication* set forth by Shannon
(1948), loosely abstracted and adapted for our purposes in figure 9.8. This
model enumerates three components to account for in the process of successful communication: the source, from which a message is broadcast by a transmitter; the destination, where a receiver decodes the message; and the channel,
which may introduce variable amounts of noise that may alter the message
at its destination. This theoretical model has been used to govern successful
message transmission in the field of telecommunications. Shannon does not
tell us how to come to meaning (Hidalgo, 2015), but coding/decoding are an
apt analog to construction/deconstruction, as discussed in part I.

Consider that the source broadcasts a particular behavior. One way to
quantify this behavior is using information, measuring, say, the number of
bits needed to describe the message. Bits correspond to how many physical
elements with an on and off state (e.g., a transistor or a light-emitting diode)
are needed to encode the source's current state. This is an objective measure
that models the complexity of the source's behavior: if the source can do
only two things, we need only one light-emitting diode, or 1 bit ($\log_2(2)=1$)
to indicate which state it is in, but if the source can do 100 things, we need
7 bits ($\log_2(100)=6.6$) and seven diodes. A model for describing not only the
internal state of a robot with bits, but also its external state, was introduced
in section 7.4 of chapter 7.

From the moment it is broadcast, this source message may then be
decoded by a receiver with a different decoding scheme than the source

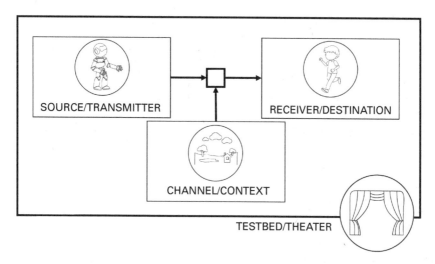

Figure 9.8
An information-theoretic model for understanding communication through movement. This model is based on the model for communication in (Shannon, 1948), which divides the process into a source, encoded by a transmitter that broadcasts across a channel (of varying "noisiness") and decoded by a receiver to get the message to a destination. As seen in this book, the performing arts is a perfect testbed to better understand this picture. This is also a useful frame for engineers to understand and systematize the methodologies used in this book.

used to encode the message, resulting in a different interpretation at the destination than was intended at the source. Moreover, the channel may introduce changes to the signal that is broadcast, again resulting in a message at the destination that is distinct from the one that was created at the source. But, when these models are aligned and dynamic error-checking schemes are successful, the message reaches its destination.

We can apply this abstract model—that has already helped develop robust mechanical and electrical systems for communication—to constrain the expected success of robots in creating and interpreting movement behavior, factoring in differences in perception across individuals as well as contextual measurements. If the source is simple and impoverished, there is a fundamental limit on how much can be communicated. On the other hand, the source can be as complex and rich as imaginable, but if the receiver does not have the correct decoding scheme, the intended message will never reach its destination. And, to extend the metaphor, humans should be modeled as destinations having their own unique, unknown, and constantly changing decoding schemes; therefore, each will make its own unique, transient meaning. These issues occur in every presentation of art. This is a core motivation that suggests to us that the arts—especially dance, with its focus on categorizing, classifying, and creating gross bodily movement profiles—constitute an essential resource of expertise for the field of robotics (Herath et al., 2016), artificial intelligence (Stark & Crawford, 2019), and human-machine interaction (Jeon et al., 2019).

In particular, the performing arts provide a natural testbed for teasing apart the complex interactions between human and machine counterparts. Jochum et al. (2016) applies this paradigm in the theater to better understand how caregiving robots might be understood by humans, positioning applied theater as an important tool in this long-term societal goal. Cuan et al. provide a model (2018) for embedding HRI experiments into live dance performance both on a proscenium stage (2019a) and in an interactive installation (2019b). Fallatah et al. (2019) discuss the challenges of handling machine failures in live performance, creating a taxonomy of four error types with distinct fixes. This type of work guides research in broader, public contexts as well (see Herath et al., 2017; Cuan et al., 2020).

9.5.1 Revisiting the Uncanny Valley

This way of viewing robots—as expressive devices whose internal states can be transmitted via a quantifiable number of bits—can reframe the commonly cited uncanny valley, in which researchers predict a dip in evaluations of likability after a certain point of lifelikeness in bodies. In the model of Mori et al. (2012), for instance, plotting likability as a function of realism, this feature (or valley) is even more prominent for agents that move, like robots, than for static sculptures or dolls. Put another way, machines in the uncanny valley have the wrong amount of "human likeness": if they were simpler devices or more convincing replications of real humans, the devices would become acceptable again. Jochum and Goldberg (2016) describe this result:

> Mori maps the relationship between affinity and human likeness on a graph, where the horizontal axis is the degree of an object's similarity to a living human and the vertical axis is the degree of affinity humans have for the object. Mori posits a non-linear function with a sharp negative extreme (loss of affinity) as likeness increases beyond a critical point (where phenomena start to appear "too close for comfort"). (p. 161)

They also describe how it is informed by art:

> This interest [in creating realistic humanoid robots] coincides with a renewed interest in mimesis and figural sculpture in the 1960s and 1970s that raised the threshold for the representational uncanny in visual art. Sculptures by George Segal (*The Dinner Table*, 1962), Frank Gallo (*Walking Nude*, 1967) and John D'Andrea (*Couple*, 1971) are human-scale statues that reproduce human anatomy in precise detail, provoking aesthetic defamiliarization that renders the human body simultaneously both familiar and unfamiliar. (p. 163)

In a broad sense, robot designers might understand the lesson as "Don't make your robot too lifelike." But in trying to understand the mechanisms at play in creating this challenge for robot designers, Jochum and Goldberg (2016, p. 165) highlight the role of the ever-evolving presentations in the visual arts in forming human impressions and opinions, creating a "shifting ground of the uncanny."

Further complicating Mori et al.'s model, we can consider the phenomenon's edge cases, remembering that the model aims to predict averages over a large subject pool. For example, the robot Sophia, built by Hanson Robotics, is a robot commonly cited as having an uncanny effect on viewers (Männistö-Funk & Sihvonen, 2018). Among other things, its eerily stretched rubbery surface that barely accommodates the motion of a mechanical hinge (meant to be re-creations of skin and a jaw) can leave a creepy impression with interactants. However, it is unlikely that David Hanson, Sophia's inventor, would consider the device creepy—he is proud of and familiar with the impressive device that he and his team have built! In fact, writing about the uncanny valley more broadly, he argues that simple (very unlifelike) robots can leave a similar impression (Hanson, 2005).

Robots judged to be "uncanny" may be seen as defying the media equation suggested by Reeves and Nass (1996). Although these devices are artificial, mediated, or symbolic representations of agency, they do not seem real. Likewise, Bartneck et al. (2020) stress the complexity of anthropomorphism, describing it as a phenomenon that relies on multiple factors, including culture. That is, the arrangement of the axes in Mori's plots trying to understand the uncanny is difficult to create in a repeatable fashion because what one person considers realistic or likable may be different for another subject.

The measure presented in this section (as well as section 7.4 in chapter 7), using bits to arrange different bodies against a single coordinate, might be a more repeatable characterization of moving platforms. Likewise, the BESST System establishes a middle ground between lay descriptions and measured quantities on which to evaluate a robot's movement: Does it have a consistent relationship to gravity (axis of length)? Does it exhibit an affinity between light motion quality and the vertical (high) dimension? Does it create a consistent sense of phrasing in its motion that allows observers to understand how many different actions it is performing?

Upon seeing the component pieces of movement, as described in the BESST System, we have found many possible interpretations for the same scene. Consequently, there is no universal measure of likability or effectiveness for any single device. So we argue that building robots with embodied movements that are meaningful to human viewers or creating systems that interpret the action of human interactants correctly are processes that will require extensive hands-on design from human engineers *and* artists.

9.5.2 Revisiting Movement Primitives

If indeed some sort of communication model can be used to understand human movement, what, then, are the symbols used in this correspondence? The typical way that researchers try to answer that question is with the notion of movement primitives (Bregler, 1997; Fanti, 2008). In chapter 1, we highlighted the pioneering work of Del Vecchio et al. (2003), who used motion primitives to describe human behavior. That chapter included the observation that "the same movement model used to generate robotic behavior through actuation can also be used to interpret the behavior of counterparts through sensing." This, then, is another way of phrasing the problem of understanding human movement, either for interpretation (as in AI, designing computer vision algorithms for human settings) or generation (as in artificial embodiment, manufacturing expressive robots): can we find a set of movement primitives that describe human movement?

This chapter has presented the usage of the BESST System to characterize movement somewhere in between "low-level" descriptions (e.g., the joint rotates 45 degrees) and "high-level" descriptions (e.g., the robot and human are friendly collaborators planning a trip together). This practice parallels progress in computer vision. For example, Saenko et al. (2012, p. 2) use "mid-level primitives" to capture "the structure of interactive verbs" and improve overall recognition of human movement. Researchers like this team and Del Vecchio's team are left to align these primitives with verbs from an English dictionary. As highlighted in the example of Luke and C3PO, however, the descriptive power and observational approach of lay human language are more contextual and personal than the BESST System's taxonomy as employed by Luo et al. (2020).

Motion primitives have also been used to generate robotic movement (Nakaoka et al., 2004; Belta et al., 2007; LaViers & Egerstedt, 2012; Lagriffoul et al., 2018). In these examples, as in the case of computer vision, there is a fuzzy line between researcher-defined primitives, based on their own somatic experiences, and mathematical descriptions, including those extracted from data. In Amy's own doctoral work, much of the design of a finite state machine came from a careful study of classical ballet and social dance forms that was both limited by her own bodily experience and also inextricably mixed with her understanding of what could be captured in a particular model. For example, warm-up exercises from the ballet *barre*, studied in a particular weekly course at the Atlanta Ballet, were simplified to create an "interesting enough" finite state machine (LaViers & Egerstedt, 2011). Disco dancing was similarly modeled, using observations of disco dance on YouTube in the early 2010s made by a person (Amy) trained in Western proscenium-based dance (not disco) using poses that were tacitly designed to read clearly on a Softbank NAO humanoid robot (LaViers & Egerstedt, 2012). We do not see how any purely objective and quantitative (e.g., mathematical) model could ever capture these aspects of experience that were embedded in the work to produce models of motion primitives. Likewise, in prior work using Labanotation as part of a scheme to specify robotic action (Abe et al., 2017), there is a large area where research and notator intelligence has to fill in the gaps in a way that is often tacit (and unacknowledged) in the work—and thus impossible to recreate.

Returning to the comparison between movement notation and music notation presented in section 2.3 of chapter 2, we need to acknowledge that the search for a set of motion primitives that explains human motion robustly still continues. This chapter aims to help illuminate why: meaning and salience in movement are not just about a mechanical structure in action but rely on many other factors. Even pitch—the mathematical description of which was not understood for hundreds of years after its invention—is an imperfect way of capturing the full phenomenon of music (Kelly, 2014). In this light, we encourage the subjective, qualitative analysis in the five case studies presented in this chapter, which use the whole BESST

System as a tool, as another approach for identifying motion primitives: the basic ideas in the Body component (e.g., basic body actions) are primitives that get modified by higher-order concepts of motion (e.g., phrasing and effort) that are represented farther up the inverted triangle shown in figure 3.4 in chapter 3 as well as the component constellations presented in this chapter.

What if we created a data classifier (see, for examples, Saenko et al. 2012; Peng et al. 2018; Wang et al. 2018) based on Lagasse's action? While researchers have made significant progress in object and movement recognition, open problems are highlighted through the lens of somatics and choreography (and revealed by our inability to robustly notate such movement). Specifically, the simplicity of extracted models and the confounding nature of context leave many open challenges in this domain, which may be helped by movement analysis.

First, note that these systems do not extract all the information about Lagasse's movement from the scene (not even all the information discussed in our simple analysis here): even state-of-the-art computer vision algorithms extract rigid, linklike skeletons (see figure 10.2 in chapter 10) that look more like robots than humans. These models ignore our fleshy viscera and rarely model breathing. Yet, as we have seen, our perception of Lagasse's breathing is part of how we identify the pattern of his movements—and relate it to our own somatic experiences. This demonstrates an important principle: the label set—or notational abstraction—that we use to describe the data impacts what can be classified.

A second problem is that Lagasse's behavior is filmed because of how *unusual* it is, not how *normal* it is. Thus, using classifiers built on his behavior will likely mischaracterize moments of cooking in real kitchens. Moreover, Lagasse's focus on a particular style of cuisine (e.g., Cajun and Creole), due to his own cultural context, creates a limited repertoire of actions: he is not folding dumplings, frying collard greens, or flattening flour into tortillas. This marginalizes communities with cooking styles that may not be prevalent—or even represented—in the training data.

The central design challenge for machines that will move with us (or observe our movements) is that unique embodied experiences create individual points of view for humans. When building a pedestrian bridge or designing a vegetable peeler, knowing the range of human weights or the widths of human hands is enough to create specifications that enable successful design. By contrast, creating movement on a robot that means the same thing to a wide variety of people requires understanding more than the physical measurements of their bodies. This design challenge requires an understanding of a myriad of complexities, including users' shared culture, distinct experiences, current environment, social context, and changing moods. In other words, designing a successful robot that functions correctly in a human-facing environment shares all the challenges of writing a best-selling novel, painting a masterpiece, and, perhaps most of all, choreographing a dance.

Chapter Summary

This chapter has presented examples of rich movement behaviors on both natural and artificial bodies. We have also used this space to consider research horizons in movement: challenging open problems and goals and how these can be reframed through the lens of movement theory. Finally, we described a model for understanding how information is transmitted through motion and contextually made into meaning by human viewers—a pair of distinct but interconnected processes that will help guide us as we work to broaden the expressive capacity of robotic systems to interpret and generate complex movement profiles. To this end, in the next chapter, we introduce notation that serves as a tool for identifying distinct movement styles and analyzing the meaning of movement.

10 Notating Movement: Advanced Analysis through Symbolic Representation

What does it mean for two distinct bodies to "do the same thing"? The answer depends on your representation—that is, on your *notational abstraction*. This chapter will compare and contrast various forms of movement representation. In so doing, it creates an opportunity for the reader to more deeply observe motion and provides tools for translating movement ideas between distinct bodies. It also reinforces the notion that training and context affect what we observe about movement.

Chapters 1 and 2 both ended with attempts at grappling with notation, and the chapters in part II introduced symbols for the taxonomy presented. In chapter 1, we painted two distinct pictures of noticing movement: one as a quantitative, objective record (section 1.2) and one as a qualitative, subjective record (section 1.3). In wrapping up the chapter (section 1.4), we offered readers an immediate, intuitive inroad to notation, kinesthetically attuning to movement through unstructured line drawings, motivating that this act of recording was a bridge between the two approaches: the line drawings were certainly subjective (being given very little structure or convention), but they also reflected objective features that were observed (like a change of direction or increase in pace). In chapter 2, we shared the plight of choreographers like JaQuel Knight, who have little ability to protect their intellectual property due to the lack of abstract, reproducible record keeping associated with human movement.

While Knight turned to the system of Labanotation as a tool in his quest to record his choreography, in this chapter we turn to *motif*, a system that aims to be more abstract than Labanotation—and therefore reproducible across bodies (as leveraged in the research described in section 5.4 of chapter 5). We even suggest that a motif of Knight's choreography could provide greater protection to his intellectual property because its abstraction applies better across distinct bodies. Thus, if an eight-legged robot was performing

Knight's work, a motif would reveal this similarity. However, as we know it today, motif is too broad and flexible to be applied in this way. Thus, we suggest a series of new conventions for the form that could aid in this task. This chapter ends with opportunities for the reader to practice notation for themselves.

10.1   Types of Movement Recording and Systematic Notation
Technology for capturing movement events and experience

Technology enables us to record aspects of movement events, but notational systems have a higher calling, requiring symbolic abstractions that record ideas with enough specificity to translate to another instance of performance. Musicians have such a tool and therefore can use a largely agreed-upon notational system to document their creations. Dancers and movement practitioners have many tools that attempt to capture movement abstractly, but none with the same level of success or widespread adoption. This is not because dancers have not been working on the problem for as long, but because the problem is much harder. The space for exploration is bigger and more complex in the human body than in a given human-built instrument. This section reviews technological and notational systems for representing human movement.

10.1.1   Recording Systems
Movement analysis led to an invention with widespread application in many creative and scientific pursuits: motion-picture recording. Eadweard Muybridge was curious about whether a horse maintained one hoof in contact with the ground throughout its gait cycle. As a horse runs faster and faster, this becomes harder to discern with certainty; so Muybridge (1887) used an existing recording method, still photography, and took many, many pictures very close together in time, resulting in a strip of images that when laid left to right (showing time progressing to the right) revealed new insights into animal gait. From there, the idea of rapidly spinning these closely shot photos in sequence, creating the illusion of movement (as in figure 10.1), evolved into what we now know as video recordings (Laurier, 2013).

Today, video remains the best and most used format for recording, sharing, and preserving choreography and other movement phenomena. Dancers use the medium to keep track of rehearsals, and biologists use it to

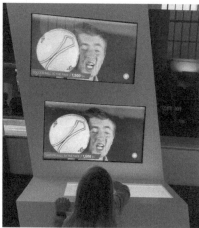

Figure 10.1
Exhibits in the Science Museum of Virginia. These exhibits explain how a series of still pictures can be used to record movement. Left: Muybridge's famous experiment, studying horse gait with rapid succession photography, is used to create a wheel that allows the participant to manipulate a series of static horse images into a sense of a running horse by spinning the wheel. Right: A participant controls the frame rate in video playback via a mounted dial; slowing the frame rate allows the participant to notice details of movement not evident in real time.

record and study animal behavior. Video has also become a rich format for the arts, with movies, television shows, and documentaries used to express all sorts of ideas native to the two-dimensional format of video. All these formats are possible since video gives us information that photos do not—namely, how the subject is moving and changing over time.

Motion-capture systems used by research labs and movie production studios (figure 1.2 in chapter 1 and figure 10.2) allow us to expand beyond the limits of two-dimensional recording, capturing a three-dimensional recording of a movement event. These systems use passive, reflective markers or active, instrumented markers worn on the body of the subject to capture its movement. Then, the systems use a redundancy of cameras or other sensors to capture movement from multiple angles for three-dimensional re-creation. Sometimes these systems are paired with traditional video to reveal further depth of detail, (e.g., Carnegie Mellon's "panoptic" studio (Joo et al., 2015)), fusing data from multiple sources; or they are used alongside the BESST System (e.g., by leveraging shape qualities to improve the

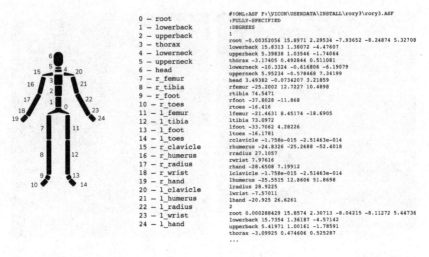

```
0 — root
1 — lowerback
2 — upperback
3 — thorax
4 — lowerneck
5 — upperneck
6 — head
7 — r_femur
8 — r_tibia
9 — r_foot
10 — r_toes
11 — l_femur
12 — l_tibia
13 — l_foot
14 — l_toes
15 — r_clavicle
16 — r_humerus
17 — r_radius
18 — r_wrist
19 — r_hand
20 — l_clavicle
21 — l_humerus
22 — l_radius
23 — l_wrist
24 — l_hand
```

```
#!OML:ASF F:\VICON\USERDATA\INSTALL\rory3\rory3.ASF
:FULLY-SPECIFIED
:DEGREES
1
root -0.00352056 15.8971 2.29534 -7.93652 -8.24874 5.32708
lowerback 15.8313 1.38072 -4.47607
upperback 5.39838 1.03546 -1.74064
thorax -3.17405 0.492844 0.511081
lowerneck -10.3324 -0.616806 -6.19079
upperneck 5.95234 -0.578468 7.34199
head 3.49382 -0.0734207 3.21859
rfemur -25.2002 12.7227 10.4898
rtibia 74.5471
rfoot -37.8028 -11.868
rtoes -16.416
lfemur -21.4631 8.45174 -18.6905
ltibia 73.0972
lfoot -33.7062 4.28226
ltoes -16.1781
rclavicle -1.758e-015 -2.51463e-014
rhumerus -24.8326 -25.2688 -52.4018
rradius 27.1057
rwrist 7.97616
rhand -28.6508 7.19912
lclavicle -1.758e-015 -2.51463e-014
lhumerus -25.5515 12.8606 51.8698
lradius 28.9225
lwrist -7.57011
lhand -20.925 26.6261
2
root 0.000288429 15.8574 2.30713 -8.04215 -8.11272 5.44736
lowerback 15.7354 1.36187 -4.57142
upperback 5.41971 1.00161 -1.78591
thorax -3.09925 0.474606 0.525287
...
```

Figure 10.2
Example of a motion-capture data structure. This data structure is used in multiple types of capture systems, including three-dimensional capture with active and passive markers (e.g., Natural Point's OptiTrack, seen in figure 1.2 of chapter 1), as well as skeleton extraction from depth cameras (e.g., Microsoft's Kinect Azure). Typical skeletons extracted have tens of degrees of freedom to model human motion, although some specialized applications use more detailed designs. Each joint (left) is given an anatomically inspired name, as listed in the skeleton file (center). Then these joints are animated with a data file (right) that determines the angles over discretized time steps with a very high frame rate. The right side of figure 8.5 in chapter 8 shows an example of motion-capture data. *Note*: this model leaves out many known joints (e.g., the vertebrae), and actual human joints are not accurately described by a simple single degree of freedom with fixed axes of rotation, as modeled in this data format.

immersive experience (Swaminathan et al., 2009)). However, even these formats are limited by the level of detail in the motion models used to explain the data. Most use a rough model that focuses on distal skeletal action, leaving out the fleshy expansion and sound of breath, the familiar twitch of skin over excited muscles, and the smell of perspiration—all of which may affect how a human notator represents the same movement—producing a highly repeatable but simplified recording system.

10.1.2 Notational Systems
Since as early as the 1700s, movement notation systems aimed at gross bodily action (as distinct from music notation which began thousands of

years prior) have been developed for various purposes. Often, notation systems are linked to musical scores and are primarily concerned with noting a singular event for the purposes of recording, preserving, and reconstructing choreography. The goal has been to document actual and specific movements for reproduction of artistic work on able-bodied human performers. For example, *Afternoon of a Faun*, a famous ballet choreographed and performed by Vaslav Nijinsky in 1912, before video recording, is often recreated from notational scores made of Nijinksy's performance of the work (Guest & Jeschke, 1991). Moreover, several types of movement notation have been used to program robot motion (Laumond & Abe, 2016). Several systems of notation are listed and briefly described in box 10.1.

As described in chapter 2, we can even consider the first forms of music notation, which were hugely important technological innovations of their time, as forms of movement notation. Tablature told musicians where to pluck instruments as a way of recording song (Kelly, 2014). This system is found on early cuneiform tablets dating to around 2,000 BC (The Schoyen Collection, 2022). This was before more abstract concepts (e.g., the neume and note, developed by monks working to record the performance of religious songs and chants in the Middle Ages) led to modern notation. The mature notation that we know today records enough information about the song that a trained expert who has never heard it can reproduce it well simply by reading the score. Moreover, it can be used across instruments, including with the human voice, and in this sense, we have described it as being *platform-invariant*.

This notation requires an abstraction away from the physical performance of the act (as the way to make a C note on any number of instruments requires a different movement) and emphasizes pitch and rhythm over things like emphasis and performance—areas where modern performers can distinguish their own versions of the work. In such a sense, we can consider modern music notation having a singular goal in sound, but a wide multiplicity of possible prescriptions in movement. Likewise, when we ask a machine to imitate a feature of human movement, we need a relaxation of how the movement can be performed due to the distinct anatomies of the movement platforms. While Labanotation and similar systems work to allow someone who has never seen a dance understand what it is, they do not afford executions on bodies of distinct morphologies in the same way (thus we can say that they are *platform-specific*). The notation relies on a

Box 10.1
Types of Movement Notation

- **Beauchamp-Feuillet:** Beauchamp-Feuillet notation was published in the early 1700s and was created to notate Baroque dances of the period so that courtiers could learn the dance sequences and perform them well when visiting the royal court. This system notates the steps of the mover along a floor pattern that the dance utilizes. In addition, the musical score is written across the top of the page and linked to the steps happening at specific points in the music.

- **Zorn:** In the mid- to late 1800s, Friedrich Zorn created a notation system, also linked to musical scores, which consisted of columns of figure drawings under each bar of music, showing the floor pattern and the progression of the dancers' form. Zorn's primary goal was to record dances that had already been made to preserve and reconstruct them.

- **Labanotation:** Labanotation was developed, initially by Rudolf Laban in the 1920s and further codified by many others, most notably Ann Hutchinson Guest (2013), to record the body moving in space. While created primarily as a means of recording choreography, it has been used to record human movement in other contexts as well—and even used to program robot motion (Abe et al., 2017). It is organized as a vertical staff, with the center line indicating weight support through the legs, and it is read from bottom to top (i.e., time moves up the page instead of left to right). Each subsequent staff line, extending right and left from the center line, organizes actions of body parts from proximal joints out to distal limbs, reflecting the bilateral symmetry of the human form. It is focused on recording already created and/or observed movement (see figure 10.3 for an example). Many Labanotated scores also reveal the progression of accompanying music alongside the movement notation. It is also used in documenting intellectual property for choreographers like JaQuel Knight, as described in chapter 2 and in (Milzoff, 2020).

- **Benesh:** Developed by Joan and Rudolf Benesh in the 1940s, this notation system was developed particularly for recording ballet choreography. The notation appears and is read similarly to music notation: that is, five lines on a staff read left to right. Each line on the staff represents a part of the body (from the floor and the bottom of the staff: foot position, knees, waist, shoulder, and head). Each "frame" (analogous to a musical note) records the dancer's positions in sequence. This structure creates compatibility between the movement and musical score, allowing Benesh notation to be written directly below and read alongside the musical accompaniment.

(*continued*)

Box 10.1 (continued)

- **Eshkol-Wachman:** In the late 1950s, a team out of Israel developed what is known as the "Eshkol-Wachman notation system." This is the first notation system that was created to be used both on paper and in the computer. Concerned primarily with joint angles and rotation, the system utilizes geometric models with a spherical system of coordinates. It has been used to study nonhuman movement and has no specific relationship to music as the previously mentioned systems do.

- **Motif:** Emerging out of Labanotation and using many overlapping symbols, motif developed inside the evolution of Laban/Bartenieff Movement Studies (LBMS), with a focus on capturing salient and essential aspects of movement and movement patterns. It can be used to create, record, and clarify movement action and intent. Many symbols have been added as active practitioners find needs for new symbols that support their particular application. It has also been used to specify robotic motion, across distinct platforms, as described in section 5.4 of chapter 5 and in (Jang Sher et al., 2019).

---

one-to-one correspondence between bodies, as used in methods that translate Labanotation scores for robotic movement (Salaris et al., 2017). This is an aspect of ableism—assuming that everybody has two arms, two legs, a wide range of motion, etc.—that limits our extant notation systems (our technology) for translating movement between distinct bodies.

In other words, we do not know (or don't agree on) what "middle C" is for moving bodies. On the one hand, motif will not give us the level of specificity and detail of the movement notation schemes described here. On the other hand, it will not require the same physical features of the body implied by these systems. Notice that the free-form wiggles introduced in section 1.4 of chapter 1, which did a surprisingly good job of recording movement, do not imply a specific platform on which the movement needs to be performed. For this reason, we have found motif to be a useful tool for translating movement ideas across bodies compared to more formalized types of notation. We will also propose some new conventions (in section 10.4) that improve the consistency with which motif may be used, attempting to bring it closer to something like the powerful abstractions of music notation.

Figure 10.3

An example of a Labanotation score. Left: The mover (Cat) starts by standing with her legs underneath her hips in a neutral stance while her arms are relaxed alongside the body. The mover then bends her knees, or *pliés*, while her arms shift. The mover then rises to *relevé* while both arms shift asymmetrically. The mover takes two steps forward, and her arms widen symmetrically and then she takes two steps to the side while her arms form an asymmetric frame around the mover. Then, the mover steps her feet together with a final flourish of her arms. Right: The score is read as if the mover is walking on it (bottom to top). It begins with place middle for the right and left legs (the lowest position of the center columns) and indicates right-low and left-low for the arms (the lowest position of the outside right and left columns, respectively). Next, the leg columns indicate place low to represent the *plié*, and the arm columns describe side-middle to place high (for the right arm) and forward-middle to place high (for the left arm). For the *relevé*, the score indicates both legs in place high, with the arms returning to side low (right and left, respectively). Next, the legs begin stepping; this breaks the phrase into two symbols three different times: first the two steps forward (right leg forward-middle, followed by left leg forward-middle in the center right and left columns, respectively), then the two steps to the side (right leg side-middle, left leg side-middle), and finally stepping together (right leg back-middle followed by left leg place middle). During the stepping pattern (the last half of the score), the arms are given various spatial directions, both symmetric and asymmetric, which is reflected in the symmetry and asymmetry of the score. The column corresponding to the right arm reads: right-forward-middle, right-side-middle, back-low, and then right-side-low. The column noting the left arm reads: left-forward-middle, place high, back-low, and then left-side-low. Here, the spatial symbols (the same ones presented in figure 5.2 in chapter 5) are analogous to musical notes. The lengthy description in this caption belies the richness of even a very simple score. At the same time, the score represents only part of what the human mover did. For example, the focus on spatial directions leaves out aspects of motion quality, demonstrating how training, taxonomy, and symbolic representation create a bias in what is observed and what can be transferred to other bodies. At the same time, the authors' training (e.g., in ballet) biases how we use the elements of the score (in both reading and writing).

10.2 Introduction to Motif

Notation for determining the essence of a phrase of movements

The symbols utilized in motif derive in large part from those used in Labanotation, but as the BESST System has evolved, new symbols have been created to reflect the evolution of the system and better capture essence in movement expression. Throughout part II, we introduced our own preferred set of symbols in the context of technology design. Books and practitioners working in different applications use slightly different symbols and conventions that we have used and also edited, leaning heavily on the symbols in Guest (2000) and Studd & Cox (2013/2020).

Motif is used to both create new movement phrases and discover new patterns, or to capture the essence of an already created movement event. The goal is to understand new movement experiences, as well as to clarify and communicate intent and expression of movement. The idea is to find what is essential, recognize patterns, establish context for, and identify the intent of the movement expression (see box 10.2 for a detailed list).

The idea of motif is not to record the details of a singular, reproducible movement event, but rather to discover patterns that express what is crucial to the event, both as perceived by the observer and experienced by the mover (noting these may be different). Thus, there is a many-to-many relationship between any given movement sequence and any given motif. This creates an opportunity for translation between bodies, for creative transformation and genesis, and for a finding of the essence (or meaning) of a particular sequence for a particular observer. That is, many, many movement sequences can be said to be described by the same motif, and many motifs can be said to describe any given movement sequence. Similar to the scribbling exercises shown in figure 1.4 in chapter 1, this creates a broader opportunity for using notation for reflection, learning, refinement, and the creative process.

10.2.1 Types of Motif

There are different styles of writing motif—or three types of *staffs* onto which symbols may be placed to create the notated score (see box 10.3 and figure 10.4). Different contexts befit different staffs depending on what needs to be revealed, communicated, and understood. The component constellations used in the case studies in chapter 9 are examples of motif as

---

Box 10.2
Uses of Motif

Notate and notice movement

- Capture, or visually represent, movement patterns and sequences, with a focus on the core essence rather than every detail
- Deepen awareness of a particular movement event
- Create a tangible, concrete artifact of the ephemeral, fleeting movement phenomenon

Construct and design movement

- Help to find, create, and iteratively design new patterns of movement, including developing a more refined physical performance of the sequence, bringing experience and expression into closer alignment
- Illustrate the contextual relationship between movement elements (i.e., foreground what is the core action and background what is a modifier)
- Clarify the relationship of parts to whole by recognizing large, overarching patterns, despite the more minute patterns within, that support the whole expression

Interpret and share movement

- Reflect choice (both of the mover and of the observer/notator of the movement)
- Assist with coming to a shared perspective, across multiple observers, in the process of observation
- Transfer the same movement sequence across distinct (human and machine) bodies

---

well; we say that they use a **triangle staff.** There are three additional styles commonly in use by Certified Movement Analysts (CMAs) today.

The **constellation motif,** which uses the **constellation staff** (see the left side of figure 10.4), is a group of "ingredients" that are present in the movement event. It reveals elements essential to the movement, but it does not reveal order, frequency, or duration. It simply notes that all these things must be present. It can be understood as the list of ingredients needed to prepare a recipe, but it is not the recipe itself. As such, all listed elements are equally important to revealing the movement event. Any symbol can be used in a constellation motif to indicate an idea that is present. The

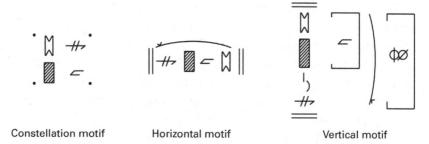

Constellation motif   Horizontal motif   Vertical motif

Figure 10.4
Three types of motif. Types of motif using symbols for rotation, high, an impulsive phrasing bow, spoke-like directional mode of shape change, bound and direct effort quality, and, in the vertical motif only, the Stability/Mobility theme. Left: a constellation motif. Center: a horizontal motif. Right: a vertical motif.

component constellations used in the case studies in chapter 9 are similar to this more traditional type of motif, adding in the structure of the BESST System outlined in figure 3.4 in chapter 3.

The **horizontal motif,** which uses the **horizontal staff** (see the center of figure 10.4), indicates essential elements and adds information about the order in which the actions occur. It can be useful when the central action is the primary expression and does not necessarily require additional details to be understood. While some modification support is possible (e.g., phrasing bows that reveal segmentation of parts that belong together inside the larger expression), it can be cumbersome in this format. The horizontal staff also allows understanding of the beginning, middle, and end, but it does not give any information about the duration of those parts of the phrase. It is written from left to right across the page, bounded by double bars on either end. As in the constellation motif, all elements are equally important, but the order in which they occur (in time) is delineated.

Finally, the **vertical motif,** which uses the **vertical staff** (see the right side of figure 10.4), reveals essential elements, the order of occurrence, and the relative duration of actions and events. It also allows the modification of primary actions (in the central column) with other things that are supportive in adding richness to the essential action (columns may be added to either the right side, the left side, or both). Most notably, in the vertical motif format, multiple modifiers and phrasing bows can be used to reveal emphasis, tone, thematic information, and more. It is written from the bottom of the page to the top.

Thus, the three types of motif commonly used in the LBMS community are listed in box 10.3 and illustrated in figure 10.4 using symbols from the BESST System introduced in part II: rotation from Body, the high spatial pull from Space, an impulsive phrasing bow from Time, spoke-like directional mode of shape change from Shape, and bound and direct effort quality (along with an added layer of thematic information, Stability/Mobility in the vertical staff).

Motif can be rudimentary or complex, general or detailed; the notator has complete control over deeming what is essential. This is what is beautiful and befitting about motif, especially for our purposes of technology development. Rather than trying to write all the details of a specific movement instance (on a specific mover), as Labanotation and similar systems do, motif encourages the notator to make choices about what is essential to the core idea of the phrase. This is true in all three formats of motif, but the inclusion of relative duration in vertical motif makes it appealing as a potential format for a detailed and sufficiently abstract system of notation for translating movement phrases between natural and artificial bodies.

Box 10.3
Types of Motif

- **Constellation motif:** The constellation staff uses the form of four dots in the corners of a square with elements contained within. It is useful for capturing the mood and tone of a cluster of movement ideas; there is no specificity of temporal relationships.
- **Horizontal motif:** The horizontal staff uses two bars at the beginning, then a sequence of symbols left to right in time, and finally two bars at the end. It is useful for capturing sequence; there is no specificity of relative duration or ability to modify the main action.
- **Vertical motif:** The vertical staff uses two bars at the bottom (the start) and the top (the end). A sequence of symbols is written up the page between these two pairs of bars. Modifying symbols may be placed to the right or left of this central column containing the "main action" (or most important ideas about the movement expression). Single bars can be used to indicate the start and end of phrases within the longer sequence. It is useful for capturing more details of a sequence; it is more laborious to create than the other two motifs.

We do not yet know how to notate movement as we notate music; that is, how to create an abstract score that communicates a dance to a dancer who has never seen it—and may not even share the same bodily configuration as the dance's choreographer. Whereas the same piece of music can be played on any number of instruments, movement scores and robotics programs tend to be platform-specific, just like early forms of music notation (like tablature) were. Somewhat relatedly, these tools bear less formality and established convention than modern music notation. In this book, we take liberty with this openness and propose some new conventions for motif (see section 10.4) that we feel better align with the corpus of theory presented here and are particularly useful in our work with machines.

10.2.2 Vertical Motif: Action Stroke and Its Replacements

Constellation and horizontal motifs can be created rather straightforwardly from the symbols given over the course of part II, but vertical motif, which offers the deepest level of detail, needs a longer discussion. The easiest way to begin practicing vertical motif is to begin with the **action stroke** (illustrated on the left side of figure 10.5). There starts the essential designation and the first step in revealing a *perceptual* pattern: the differentiation between *action* and *no action*. Action is indicated by a line and no action is indicated by empty space. These simple notations offer a great deal of information about the movement event by identifying the number of actions and by segmenting how they unfold over time in a given movement event.

When writing vertical motifs, the option to replace and modify the action stroke with more specific symbols is available. Action strokes can be replaced by or modified by the set of symbols that were introduced throughout part II of this text. Themes, as well as phrasing, can be represented symbolically, but these symbols do not replace an action stroke; rather, they modify it (as they are not actions unto themselves). These concepts are listed in box 10.4, and their spatial arrangement in a vertical motif is diagrammed in the center of figure 10.5.

For example, if the main action is primarily about the space of forward high, that symbol would replace the action stroke. Further, it might be important to recognize that space is being revealed with a great sense of delicacy, which we could decide to notate as light weight effort, in which case an effort symbol modifier could be added. The opposite is also the case: if the primary action is about an expression (or a series of expressions) of

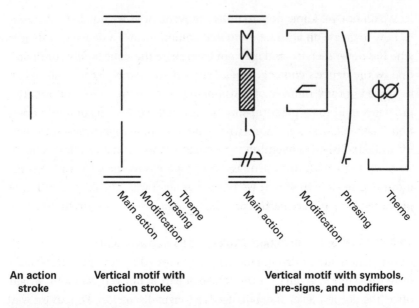

An action   Vertical motif with   Vertical motif with symbols,
stroke   action stroke   pre-signs, and modifiers

Figure 10.5
Examples of a vertical motif using only action strokes and using additional symbols and modifiers. Left: An action stroke. Center: An action stroke motif with the anatomy of "lanes" labeled: main action, modification, phrasing, and theme. Right: A more complex vertical motif using symbols from the taxonomy in part II. Symbols that elongate (high and rotate, in this example) replace action strokes, while symbols that do not elongate (spoke-like directional shape change) are used as pre-signs for action strokes. This motif also uses modifiers (bound and direct effort in this example), phrasing (impulsive phrasing bow), and theme (Stability/Mobility).

effort that unfolds in the space of forward high, then the effort symbols would be in the central column and the space symbols would modify it. This model follows the tetrahedral model of BESST, where every component of the system can be seen as the central basis for viewing motion.

This way of approaching vertical motif writing adopts the idea that the central column of the motif represents what is most important, or salient, to the understanding of the movement expression. The adjacent columns then reveal aspects that support the main expression and help illuminate layers of information to enhance the richness of the representation. In this approach, the effort and shape symbols (as well as certain other symbols that do not "elongate" to reveal relative duration) are **pre-signs** that are attached to an action stroke when they are used in the central column in a vertical motif, but stand alone in a horizontal or constellation motif. The

---

Box 10.4

Rules for Motif on the Vertical Staff

Symbols that replace action strokes

- Time component
  - ○ Relative duration (embedded in the length of a symbol in a vertical motif)
- Symbols that expand/elongate to reveal relative duration
  - ○ Body component
    - ▪ Basic body actions
      - • Posture
      - • Gesture
      - • Locomotion
      - • Rotation
      - • Condense
      - • Expand
      - • Hold
      - • Jump
  - ○ Space component
    - ▪ Spatial direction
  - ○ Shape component
    - ▪ Primary shape patterns
      - ▪ Gather/scatter
- Symbols that are used as a pre-sign and need an action stroke to reveal relative duration
  - ○ Body component
    - ▪ Basic body actions
      - • Change of support
      - • Vocalize
      - • Focus
      - • Touch
    - ▪ Body parts[2]

(*continued*)

Box 10.4 (continued)

- ∘ Space component
  - ▪ Level
  - ▪ Zone
  - ▪ Reach space
  - ▪ Pathway
- ∘ Effort component
  - ▪ All effort symbols are a pre-sign and need an action stroke to indicate duration.[3]
- ∘ Shape component
  - ▪ Primary shape patterns
    - • Concave/convex
  - ▪ Still shape forms
  - ▪ Modes of shape change
  - ▪ Shape quality

Symbols that modify action strokes

- • Any symbol that can replace an action stroke
- • Body component
  - ∘ Patterns of body organization
- • Time component
  - ∘ Phrasing
- • Thematic duality

length of the action stroke it is connected to reveals the relative duration of this symbol.[1] These concepts are summarized in box 10.4, and an example is provided on the right side of figure 10.5.

10.3   What Is Up with Imitation?
A framework for translation—and, therefore, communication—between distinct bodies

How is it that any character that is unlike a human can be expressive? We have claimed that not all robots can express the exact same physical actions as others, noting the comparison between a humanoid, with its armlike extensions enabling gestures with low dynamic constraint, and a quadruped, with a dynamic requirement for stability that at least three extensions (or "limbs") must contact the ground for static stability. In this section, we will turn that idea on its head and note how unlike bodies can imitate one another—even very, very simple movement systems can seem to imitate more complex ones from a *perceptual* point of view. In fact, having a symbolic system for notating movement becomes a point of transfer for how ideas on one body can be expressed on another—and thus a starting point for design.

Several notable examples provide evidence that this sort of abstract imitation is possible. The perception of consistent narrative in the experiments in Heider and Simmel (1944) suggest that human actors can replace (or be replaced by) abstract shapes used to create a narrative cartoon—results that surprised researchers in the consistency of the interpretation by human observers. Figure 4.2 in chapter 4 also shows examples of a variety of forms that have been used in expressive contexts, often alongside humans, to communicate consistent ideas (as human actors might). These examples show how shared experience (e.g., having the same social upbringing), can produce similar responses to the same stimulus, but they should be contextualized by the common act of comparing thoughts and opinions about art, beautiful vistas in nature, or other stimuli. For example, after watching a film, moviegoers will compare their perceptions of the work, often disagreeing about its quality or message.

In representing motion of and with machines, we can see similar feats. For example, if you ask someone to imitate an airplane, the response will likely be to hold out both arms horizontally to evoke the wings of the plane. This consistent use of our bilateral symmetry and alignment to direction of travel can be explained through the Body component of the BESST System. This same idea of imitation has been posed in reverse (a robot imitating a human instead of a human imitating a machine) in human-robot interaction (HRI) research (Jang Sher et al., 2019; Kaushik & LaViers, 2019), where human subjects were asked to determine whether a robot was doing the same thing as a human. Kaushik et al. (2018) imitated a human motion-capture skeleton using a metric derived from analyzing human motion alongside the contemporary dancer and scholar Ilya Vidrin. In this case, the device's motion was driven via a lower-dimensional signal derived from the motion-capture skeleton, and the other was driven via an uncorrelated signal. In this case, researchers could actually measure the ability of humans to detect which robot was imitating the human. For the robot with the lower number of degrees of freedom (and thus the less expressive one as measured in figure 7.5 in chapter 7), subjects could not clearly discern an imitating robot; the imitation of the robot with more degrees of freedom was preferred by subjects over three-fourths of the time (Kaushik & LaViers, 2019).

Psychologists have found that human behavior uses notions of symmetry and imitation in conversation; subjects rated conversations with digitized human conversation partners as being less effective when researchers interrupted the natural symmetry-forming movement of the partners' heads (Ashenfelter et al., 2009). Likewise, notions of imitation drove the validation of research presented in sections 4.5 in chapter 4 and 5.4 in chapter 5. Of course, none of these examples consist of perfect imitation or translation of movement from one body to another. This suggests that there are abstract concepts of movement that each body is successfully portraying. The BESST System provides one lens through which to define such abstract similarity and, as a result, gains a foothold into human perception of imitation, which seems to be an important basis (or baseline) of how we communicate through motion.

Further, we note that creating most of these examples is the result of the work of performing artists, who possess a skill set for *consciously* noticing subtle features of movement. That means that they notice movement that others do not and can articulate how to edit it for improvement toward a goal. For example, comedians who are skilled at impressions notice subtle facial and vocal features and can articulate these through their own moving bodies. The rest of us can appreciate the fidelity of the impression, but we do not understand how to create that moment. That is, we do not all notice movement with the same granularity. One of the key goals of this book—and the use of notation—is to provide an inroad into observing with more granularity.

The examples presented in chapter 9 also rely on a notion of imitation—specifically of machines imitating humans. So far, we have had a lot of fun thinking about this idea—and seen it working (and not working) in many places—but we have not given much careful thought yet to *why* machines should model human movement, either for generation or interpretation, actuation or sensing.

On the one hand, many people are interested in the task of imitating, even re-creating, human behavior in artificial systems. On the other hand, a possibly more pragmatic answer is simply to build better tools. In some cases, mimicking aspects of the human condition is necessary to build better tools. Take, for example, the OXO GoodGrips line of kitchen tools, a perennial favorite example of human-centered design that mimics the curves and soft materiality of the human hands that will use the tools. Thus, imitation may be thought of as a much simpler act of creating a tool that blends in well with human counterparts, becoming more useful and more harmonious through that imitation.

The latter approach is the stance from which we have written this book, but even if your interest is in the former, many of the same questions arise; in particular: What does it mean for two distinct bodies to "do the same thing"? Often, as the examples here have demonstrated, this question is not pondered quite long enough. Through the lens of movement studies, we have seen how a backflipping Atlas robot reveals more about new strategies for backflips than it does about humanlike motion.

Indeed, if we represent a backflip simply as a rotation of the whole body in the air, many entities do "backflips": quarters in a coin toss, pancakes on a restaurant griddle, Atlas the robot, kids jumping off a swimming pool diving board, or Simone Biles in the Olympics. If, however, we use a more nuanced representation—considering, for example, whether the flip was initiated from the core of the body or a distal joint—we immediately eliminate some of our backflip candidates: for one thing, all of the inanimate objects; for another, all of the inhuman objects; and possibly in addition, any humans using particularly odd or unorthodox backflip strategies, like the kids casually playing at the pool. Indeed, as we increase the specificity of our abstraction, we expect that only professional gymnasts, like Simone Biles, would meet the criteria. Thus, imitation depends on our representation of the action.

Consider the example of the two movement sequences pictured in figure 10.6. How many "movements" (or phrases) does each sequence convey? What elements of the BESST System are present in each sequence? *Are these people "doing the same thing"?*

The way that we notate each sequence, choosing action stroke lengths and/or replacing action strokes with more specific movement ideas, changes

Figure 10.6
Two sequences of movement presented for analysis. Top: sequence A performed by Amy, running left to right. Bottom: sequence B performed by Cat, running left to right.

whether we identify the sequences as the same or not. Certainly they have some degree of overlap: both begin standing and end crouched low, for example. But say that we notated one as four movements and the other as three, lumping together the opening head roll with the arm action; then we end up with very different ideas about the sequences. Figures 10.7 and 10.8 offer one such analysis that utilizes the same motif to notate the movement, providing a clear description of *in what way* these two sequences are the same. A different motif—for example, one that emphasizes the upward focus used by the performer of sequence B in the third movement—would result in a different answer: "No, they are not the same."

In examining sequences A and B as presented in figures 10.7 and 10.8, respectively, we would say these performers are doing the same thing. Despite some differences in the exact movement that each one performs, both satisfy the motif of the notated four movements in the phrase. Cat does an arc-like directional mode of shape change with the head swoop, while Amy does it with her right arm, and yet both are actions of bridging to the environment. Likewise, the condensing movement occurs in different spatial zones for each performer, but in both cases, they are pulling in toward themselves—both condensing their bodies. The Effort moment of the "flick" (light, indirect, and sudden effort) is accomplished by both performers with their hands—one snapping her fingers and one spreading her fingers. What is interesting to note here is that both do the action in the zone of up.

In chapter 8, we introduced the idea of affinities between components. In this particular moment, we see the affinity of Effort and Space. In this affinity, the expression of movement quality can be enhanced by a particular spatial pull. The affinity between light weight effort and the space of

Filmstrip of movement	Motif	Brief description and taxonomic labeling
	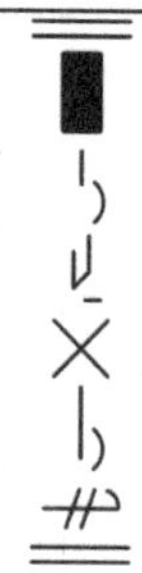	Right foot steps forward into low lunge. Low (Spatial Pull / **Space**)
		Fingers on the right handsnap (quickly). Flick (Basic Effort Action / **Effort**)
		Arms gather in front of the body in the horizontal plane. Condense (Basic Body Action / **Body**)
		Right arm reaches out, forward of the body. Arc-like Directional (Mode of Shape Change / **Shape**)

Figure 10.7

Sequence A decoded (one possible solution). Amy performs her interpretation of the motif in which the phrase is identified as having four movements. Seven photos to the left of the motif reveal the execution of the movement actions of Shape, Body, Effort, and Space (specifically delineated in the right column). The motif is read from bottom to top.

Filmstrip of movement	Motif	Brief description and taxonomic labeling
		Whole body crouches low. Low (Spatial Pull / **Space**)
		Fingers and both hands extend open, upward (quickly). Flick (Basic Effort Action / **Effort**)
		Arms gather in front of the body in the vertical plane. Condense (Basic Body Action / **Body**)
		Head swoops from left to right. Arc-like Directional (Mode of Shape Change / **Shape**)

Figure 10.8

Sequence B decoded using the same motif as in figure 10.7. Cat performs her interpretation of the same motif in which the phrase is identified as having four movements. Seven photos to the left of the motif reveal the execution of the movement actions of Shape, Body, Effort, and Space (specifically delineated in the right column). The motif is read from bottom to top.

up is seen here in the resonance of both performers choosing to express the dynamic quality in the same space. At the end, both move toward the spatial pull of low; so in terms of this (notated) aspect of the Space component, they are again *doing the same thing*. Through the (un-notated) lens of the Body component, however, one is lunging and the other is crouching. In this way, it is up to the notator to decide what BESST System element—or notational abstraction—is important for defining the expression in their context.

Can any robot replicate these sequences? Can any body, having an ever-so-slightly or maybe drastically different morphology, replicate these sequences? What does it mean to *do the same thing?* Looking through the lens of the five components introduced in part II, the answer is both complexified and enabled. Any mechanized device can be seen as a body moving in space and time. Thus, many of the ideas in chapters 4–6 can be observed in machines in motion: the physical elements of the platform, while dissimilar from human bodies, have core, proximal, and distal elements; the body may move to legible directions in space; and the actions will likely have a perceived duration.

However, higher-order ideas in these components, such as cross-lateral patterns of body organization, near-reach space, and phrasing may not be evident from the machine's behavior—either because it is such a foreign object or because it has such a simplistic movement life. Moreover, ideas of relationship and intent, dealt with throughout the system, though primarily in Shape and Effort (chapters 7 and 8), may also not be manifest—and certainly not for all viewers. For example, a person who does not know a device's purpose may not note anything of interest in a particular movement, while the designers and researchers who work with the device and have explicit knowledge of the programming and mechanisms in the device may perceive intent when they see the machine attempting a particular task because they understand the strategies at play.

Just as comparing babies to adult movers has long clarified aspects of movement analysis, such as patterns of body organization—which newborns begin exhibiting in later and later stages of development—comparing rich, expressive adult movers to machines reveals insights into how people perceive movement. We require spatial extent to see Shape; we require temporal extent to see Effort. Thus, we may think of these categories as higher-order, or more meaningful ideas about movement that manifest in the aggregate of a body moving in space and time. Likewise, we establish a sense of a concept like near and far reach through long-term observations of a mover's behavior—which we may not always have access to for new bodies, such as novel robots. Phrasing is another complex idea that requires a sense, usually, of condensing effort qualities, which is not always developed in robotic movement.

In general, then, it may not always be possible to reasonably imitate the motif used in figures 10.7 and 10.8 on a robot. Take, for example, a mobile robot like the Sphero Mini, which does not have any articulated appendages. It will be difficult to create a sense of arc-like directional shape change on such a platform. To reflect this fact, the next section introduces a stricter, more structured version of motif that may be more useful in working to translate ideas between bodies of distinct morphology (like humans and robots). This suggested staff aims to provide more information about a designer's goal. That is, we need more structure as to what the core "notes" of movement are, so we can translate a particular "song" (or dance) to another "instrument" (or moving body).

10.4 Suggesting Stricter Conventions for Motif with Machines
Notation for developing expressive machines

We have touted the benefit of working with motif when working with machines: having the flexibility of many movement patterns that can reflect the same idea creates space for translation to many distinct bodies. However, we have found that additional structure in the notation may help preserve the idea being communicated with greater specificity without requiring a specific bodily morphology. Following the idea of actions being verbs and modifiers being adverbs, discussed in chapter 9, we suggest a new format for vertical motif, diagrammed in figure 10.9 and called the **CPMMPT staff.**

Specifically, an approach to using vertical motif that we find useful is to use the central column to indicate the foundational aspects of movement (i.e., basic body actions, spatial pulls, relative duration) and the adjacent columns to the right to reflect higher-order components (e.g., patterns of body organization, reach space, phrasing, shape qualities, effort, themes). The adjacent columns to the left of the central column are used to describe more measurable aspects of movement (e.g., body parts and absolute

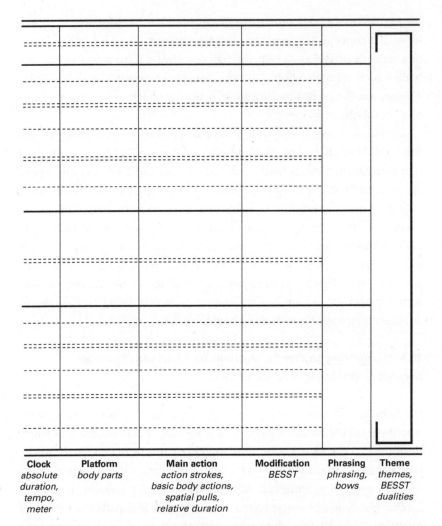

Clock	Platform	Main action	Modification	Phrasing	Theme
absolute duration, tempo, meter	*body parts*	*action strokes, basic body actions, spatial pulls, relative duration*	*BESST*	*phrasing, bows*	*themes, BESST dualities*

Figure 10.9
Our suggested staff for using motif with machines. We call this the CPMMPT staff by forming an acronym of each column name (Clock, Platform, Main action, Modifiers, Phrasing, and Theme). It is used in concert with the conventions outlined in box 10.5.

duration) which are also specific to a given physical morphology of a moving platform. The columns then shift left to right from most "measurable" to most "interpretive": in other words, they move from body toward meaning. The motif is opened and closed with a double solid line, as is tradition, but we introduce a horizontal double dotted line to demarcate where an action begins and ends and a horizontal single dotted line to demarcate where a modification, phrasing, or theme begins and ends (which may be in the middle of a particular main action).

In the convention introduced in section 10.3, all these symbols could be used in the central column or as modifiers except for phrasing and theme, which are usually reserved for the far right of the motif. This has the advantage of a simpler, less crowded, and more flexible format. Our more formalized approach utilizes added conventions, symbols, and space that may go unused in some cases.

We suggest our more tedious conventions because we find that effort, for example, can optionally modify an action with spatial specificity or temporal specificity, especially in cases of a highly expressive body, but there is always a "body" moving in "space" and "time." As such, ideas about effort, shape, body parts (and more), are always modifying actions in space and time. This empty space can be revealing, just as the component constellations in chapter 9 revealed patterns through empty component areas; for example, Emeril Lagasse's lack of use of the Space component revealed his evocative approach to spicing his culinary creations. This approach is especially useful in robotics, where the body moving in space and time may not have the manifest, complex interactions with the environment required to reveal intent, relationships, and motivation. Likewise, the expository nature of our approach aims at creating a more regular format for this notation system that can be deployed in a wider array of applications.

We have said that bodies moving through space and time are foundational to our experience of observing movement. That is, we see aspects of Body, Space, and Time—specifically, basic body actions, spatial directions, and relative duration—in all moving bodies (even very simple robots). For example, the Sphero Mini platform, which is just a rolling sphere, still has rotation, forward/back and left/right, and the ability to change the duration and rhythm of its action. More complex ideas about movement, like patterns of body organization, reach space, phrasing, shape quality, and effort quality, often require articulated bodies. For example, a humanoid has a more complex configuration space than a fixed sphere has, where more complex ideas about the device's relationship to its environment can be specified with salience for a human viewer. Moreover, such nuanced variability of motion is not necessarily something that comes naturally to human bodies; it is a skill that is practiced over years, as described in books for performing artists (Newlove, 2007; Bloom et al., 2017).

In other words, the motif's central line of actions of relative duration can be generic actions, actions with bodily specificity, or actions with spatial specificity. Higher-order ideas about movement may or may not be present. We want to formalize this concept in the structure and convention of motif and to encourage the notator to define a baseline set of actions that is easily achievable, even on a robot. Thus, the presentation that follows may seem restrictive to those experienced with motif.

Our structure (diagrammed in figure 10.9 with an instantiated example shown in figure 10.10) is as follows: We restrict the central action line of the vertical motif to be action stroke, basic body action, or spatial direction. To the left of this action line, we allow platform- and event-specific elements like body parts, absolute duration, tempo, and meter. These elements may differ based on physical platforms of various morphologies and dynamic capabilities. To the right of this action line, we add increasingly complex levels of analysis, building toward meaning. These are as follows (moving to the right): more complex elements (compared to basic body action, spatial pulls, relative duration, and tempo) of Body, Space, and Time, as well as all elements of Shape and Effort, in one column; phrasing in another; and themes in the farthest column to the right.

Box 10.5 provides the delineation of what does and does not replace an action stroke. The goal of these conventions is to clarify the readability and relationship of foundational components (Body, Space, Time) to higher-order components (Shape, Effort, themes), which become important to recognize when working with machines or across bodies of various levels of skill. We can then become cognizant of how we assign intent and motivation to artificial agents moving in an environment (and how we come to understand our own movement intentionality). The level of complexity and detail in the motif is still determined by the goal of the mover and/or observer, remembering the things that motif can do and be used for: reflection, learning, refinement, and creative process. We preserve this flexibility because it is especially useful for transferring movement ideas between bodies, representing essence, and coming to observational consensus.

To use this format of motif to notate machine movement, consider the motion pictured in figure 10.11. A small, spherical robot (the Sphero Mini) moves in circles. Each circle begins with a kick as the motor turns on

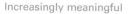
Increasingly meaningful

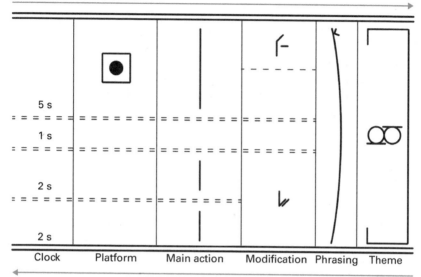

Increasingly measurable

Figure 10.10

Example of our suggested staff for using motif with machines. This motif (in black) shows three actions occurring over roughly 10 s. The first two actions use rising shape quality; the third action is performed with the pelvis. The entire sequence is an impactive phrase that explores the theme of Function/Expression. The structure of this motif (in gray) reflects the hierarchical, inverted triangle model of BESST, delineating more measurable elements (thus, ones that are easier to implement on machines) specific to a particular platform to the left of the main action and more meaningful elements (contextual products of rich bodily articulation and coordination), that are transferable to other bodies, to the right.

and slowly decreases in length (for three cycles), finishing in a slower and smaller arcing action.

This highlights the main idea about platform-invariant notation: the notation of the movement is different than the scheme that generates it on a particular body. In robotics, this means that the code used to generate a movement expression, which may be viewed as a kind of tablature, or platform-specific notation, is not the same as a motif of that expression. Imagine a program that commands thirty repetitions of the same action. After viewing this action over and over and over, the action may appear different to a human observer. Thus, in notating this movement, we needed to

Box 10.5

Rules for Motif on the CPMMPT Staff

Conventions for vertical divisions

- **Clock column** (leftmost column): optionally indicates the absolute duration, meter, or tempo of movements and phrases
- **Platform column** (second column from the left): optionally indicates a body part for a specific platform, allowing different bodies to intentionally create new assignments to accommodate different morphologies
- **Main action column** (left central column): indicates the main action, most essential to the movement sequence, using basic elements from Body, Space, and Time:
  - Basic body actions
  - Spatial pulls
  - Relative duration (e.g., as embedded in the length of an action stroke)
- **Modification column** (right-central column): indicates modifiers for the main action that offer additional information from all categories of the BESST System, excluding thematic dualities
- **Phrase column** (second column from the right): optionally indicates the types of phrasing in the sequence using phrasing bows from the Time component
- **Theme column** (rightmost column): optionally indicates the theme, or larger idea, of the movement sequence from all categories of the BESST System, including thematic dualities

Conventions for horizontal divisions

- Double solid line to start and end the motif
- Single solid line to indicate a new phrase (extends from the Clock to the Modification column)
- Double dotted line to indicate a new action (extends across the Main action column, optionally bisecting Clock, Platform, and Modification columns)
- Single dotted line to indicate any separation within a movement of entries in the Clock, Platform, or Modification columns

Conventions for elongating symbols

- Symbols in the Platform, Main action, Modification, and Theme columns do not need to elongate because the double dotted line indicates relative duration of each element of the motif.
- In the Main action column, action strokes may be replaced by elongated symbols of basic body actions or spatial pulls—or symbol readability can be preserved, letting the double dotted line indicate their temporal extent.
- Phrasing and theme bows in the Phrasing and Theme columns extend across the movement or movements used in the phrase, elongating as needed.

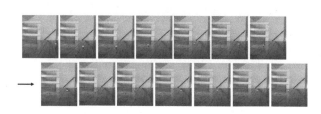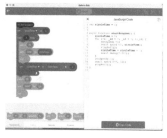

Figure 10.11

Robot motion for motif and translation exercise. The motion of a mobile robot (left, shown left to right in two rows) and the code that generated it (right).

(somewhat counterintuitively) ignore the robot command structure in order to experience the motion writ for a human observer in the environment.

In creating a motif for the robot's movement, we first used our bodies to embody it, trying it ourselves (as shown in figure 10.12). Here, we have three options for how we might imitate the movement of the machine, which help us determine how to notate it. In the first example (the top row of figure 10.12), the rolling of the robot is embodied as a physical rolling of human joints, focusing on the shifting weight and meditative quality the repetition imparts to the sequence; this execution lacks the pronounced floor patterns produced by the machine. In the second (the middle row of figure 10.12), the physical mapping between bodies is more encompassing: rolling the entire body of the robot equals rolling the entire body of the human; this execution is difficult for the human performer, resulting in a more jagged execution as the body folds and unfolds itself to accommodate each roll. Finally, in the third embodiment (the bottom row of figure 10.12), the focus is on the idea of traveling in a mode native to the body (for the robot, this is rolling; for the human, this is walking); now the resulting motion reflects an easy, organic travel to the right, in diminishing cycles, which befits the natural dynamics of Amy's body. However, this motion varies greatly between Amy and the robot (e.g., the still shape form of Amy has become more pin-like instead of ball-like).

All three embodiments are valid *choreographic choices* that result in three motifs, as illustrated in figure 10.13. Notice how the process described here requires us to move in our own bodies to see the myriad elements contained within a simple sequence of movements of a very simple device. If

Figure 10.12
Three ways of embodying a robot motion. Three imitations of the robot motion in figure 10.11, performed by Amy in the process of notating the movement. Each embodiment, presented as images to be read left to right in a distinct row, requires a different motif.

we had different bodies or different training, that process of embodiment would display different choices. Further, while the convention suggested here is stricter than traditional motif, it does not become as rigid as traditional Labanotation, which inherently requires a "humanoid" structure to translate the instructions as it is embedded in the structure of the notational staff. Thus, the CPMMPT staff preserves motif as a flexible notational structure that can describe movement on many bodies, while creating a prioritization between simple ideas of movement (e.g., action or no action) and more complex manifestations of it (e.g., effort quality).

From the examples provided in this section, we hope that you will see that notational abstractions offer the opportunity to reconcile quantitative or qualitative and objective or subjective aspects of movement. We can count (three distinct actions as reflected by the structure of action strokes, say) and describe (easy, natural gait as reflected by a modifier of free flow, say) to create impressions that objectively last the same amount of time and subjectively (according to us!) give a similar sense of movement. In this regard, the symbols on the proposed CPMMPT staff become like notes on a treble or bass clef—able to be played on instruments with different capacities for motion. Using motif in the design process of translation between robot and human (and vice versa) allows the explication of these design choices and supports the ability to create harmonious interactions and interfaces between the two. One way to conceive of this process is outlined in box 10.6.

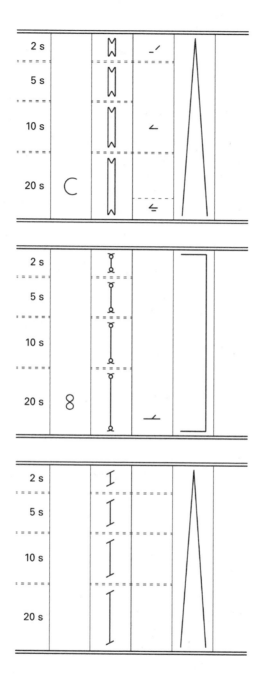

Figure 10.13

Three motifs corresponding to different embodiments. Motifs of the robot motion in figure 10.11, corresponding to the three embodiments in figure 10.12. From top to bottom these are motifs in the CPMMPT staff corresponding to: rolling the joints, rolling the whole body, and traveling.

Box 10.6
A Machine Design Process Supported by Movement Notation

You will use this process to come to consensus between an external artifact—it could be choreography of a human or robot body, hardware design, or a scheme for interaction—and your own design goals using motif. Follow the numbered steps, iterating as instructed until you've clarified your intended "machine design."

1. Begin with a record of a movement phrase. For example, you could have:

   ◦ a video recording of yourself doing an expression of interest;

   ◦ an interface with a series of buttons and instructions on how to use it;

   ◦ a sketch of the behavior of a human interactant that a robot or computer vision system will be measuring;

      or

   ◦ a series of commands for a robotic system.

2. Enact this score in your own body (this is akin to the idea of "bodystorming") or on your artificial system. This is your first draft.

3. 🎞 Record this first draft so that you can observe it, ideally with the ability to replay, rewind, and review at different speeds. (Video works well for this!)

4. Create a motif.

   ◦ If using the triangle staff:

      i. Note your context and purpose in the spaces designated.

      ii. Write down the concepts that immediately jump out at you in your first few observations.

      iii. Place these concepts inside their associated components.

      iv. Check the density of each component bin on the staff. Especially high density in one component may reveal the importance of that lens for this context or your own overreliance on that lens; likewise, empty bins may reveal an important pattern—or your own failure to consider that lens for analysis.

      v. Revisit the movement and consider rebalancing, editing, or adding to your earlier impression, repeating this step until you converge on a fitting motif and machine design.

   ◦ If using the CPMMPT staff:

      i. Note your context and purpose next to the motif.

      ii. Starting with the Main Action column, use action strokes to decide how many movements exist in the movement phrase under consideration, noting their relative durations. Then, after a few more observations, consider replacing some of these action strokes with symbols

(continued)

Box 10.6 (continued)

for a basic body action or spatial pull as appropriate. Use these symbols to create a pattern of distinct actions and phrases, using double-dotted lines to segment actions and solid lines to segment phrases.

iii. Move outward to the platform and modifier columns, adding in specificity about body part in the Platform column and more complex notions of Body, Effort, Space, Shape, and Time in the Modifier column.

iv. Move outward to the Clock and Theme columns, adding in specificity about absolute duration, tempo, and meter and any overarching themes you think are important to this design.

v. Check the density of each column on the staff. Especially high density in one column may reveal the importance of that lens for this context or your own overreliance on that lens; likewise, empty columns may reveal an important pattern—or your own failure to consider that lens for analysis.

vi. Revisit the movement[4] and consider rebalancing, editing, or adding to your earlier impression, repeating this entire sequence until you converge on a fitting motif and machine design.

5. Giving the motif that you are using to describe your design (be it for movement on either a human or robot form) to another designer and asking them to move it themselves will highlight different solutions inside your design space. Often, this clarifies essence. If you have this resource, at various points inside your iteration with your design, give your motif (without too much additional context) to a partner for this type of feedback. This can also help identify contexts or edge cases that you had not yet considered. (If you do not have such a partner, returning to your own design after a night of sleep or a few days away, can simulate this resource too.)

6. At the end of this process, you should have a motif and a machine design that reflect each other. The motif should capture abstract aspects of movement that are important to your work and help you highlight and defend the design choices you made. To return to our examples in the first step, you could have:

○ helped the movement expression in your video recording become more expressive by giving it greater clarity in terms of the choreographic choices being made and better aligning it around a central idea—and as a result, you can now more easily perform the expression consistently in your body;

○ altered the interface to give it a different set of buttons (for example) with instructions that better align to a particular use case, communicating how to use it with more nuance;

(continued)

Box 10.6 (continued)

- ○ given the robot or computer vision system a clearer (likely narrower) set
  of operating conditions where it can succeed; you are also likely better
  able now to communicate its strengths and weaknesses;
  or
- ○ edited the robot commands to create behavior that better achieves your
  design goals and, simultaneously, clarified your design goals themselves.

10.5   Exercises in Recording and Notating Movement

Practice, practice, practice

In the following exercises, you will be invited to explore ways in which
working from different means of movement generation, refinement, and
interpretation support your understanding of how notating movement—
and moving from notation—can help to clarify movement design.

- **Noticing what video misses:** In this exercise, you will explicate the miss-
  ing features of movement in a video you create.
  - ○ Create a movement sequence, practicing it until you know it and can
    repeat it consistently.
  - ○ 📷 Use a camera to record your movement sequence.
  - ○ Replay your newly created record and create a list of features that you
    know occurred but are not evident on the video. For example:
    - ▪ If you are facing the camera, it may be hard to discern the curve of
      your neck or slope of your shoulders.
    - ▪ If you turn away from the camera, it may be impossible to see your
      facial expression.
    - ▪ If one arm moves behind your back, can you tell how your fingers
      are positioned?
  - ○ Reflect on your list and movements, comparing the model of the
    movements in your head to the record of the video camera.
  - ○ Try this with a friend, who (not knowing the original sequence, will
    not be able to fill in any detail with prior knowledge of the event as
    you do) may find more holes in the video record.

Notice how your record of your movement, the recording of the movement, and your friend's interpretation of the movement are similar and different. Was your friend able to replicate your movement as you intended it to be performed from the recording?

- **Exploring motif writing:** In this exercise, you will begin to notate movement you observe (and subsequently generate) and recognize the choices you make regarding what is important to note about the movement and how you re-create those features with your own movement sequence.

  1. Take a walk and observe the movement (or movements) that you perceive in the environment.

  2. Write (in your own way) what you saw (use verbs and adverbs here).

  3. From that record, create a movement sequence that reveals what you wrote down.

  4. 🎦 Embody the movement sequence so that it is familiar and replicable—such that you can easily repeat it in your own body without stopping to remember what comes next. Now, record yourself moving the sequence and then set that recording aside.

  5. Next, create a constellation motif by distilling (from your descriptors) what elements are essential to the movement sequence.

  6. Now, take those elements and order them in a horizontal motif in which the sequence of the elements unfolding reveals the intent of your movement sequence.

  7. Finally, using action strokes, create a vertical motif that reveals action/no action, as well as the relative duration of the actions. Which action strokes would be served by being replaced with another symbol? Which action strokes would be served by adding a modifier? What does this reveal about the movement sequence and your analysis of it?

  8. Take a look at your recording and make a motif of what you observe yourself doing in that recording. Which form of motif did you jump to use? Why? What do you deem essential to communicate/explicate in order to reveal the essence of your sequence?

     ◦ 👫 This exercise works even better if you swap recordings with a partner and (if possible) work in groups to show how different people have different points of view about the movement.

Is any one form of motif more useful to you? What did you experience as different from/the same as writing a motif from your embodiment of your sequence versus writing a motif from observing yourself moving in the recording? Were new elements revealed as essential to observing versus moving?

- **Creating a vertical motif, replacing action strokes and adding modifiers:** In this exercise, you will practice creating a vertical motif of your own movement, following the conventions outlined previously. Begin by returning to the movement sequence that you created in the first exercise in this chapter, or create a new sequence now.

  1. Move your sequence. How many *actions* are there? What are their relative durations? Using only action strokes, create a vertical motif that reveals the number of actions and their relative duration in the central column.

  2. In the column immediately to the left of this, determine whether it is useful and/or necessary to specify particular body parts that are moving. If so, using the symbols in chapter 4, identify body parts moving in relationship to any or all action strokes.

  3. In the column to the left of that, note the absolute duration of the entire event (i.e., the number of seconds and/or minutes over which the entire event unfolds).

  4. Now, go back to the action strokes. For each, determine if it is mostly about a specific body action, or a specific spatial pull, or the length of time of the action. Replace the action strokes with the appropriate symbols for body actions and/or spatial directions; if neither is salient, leave the action stroke in place to indicate that something happens for that relative duration (that is not determinative of body action or space).

  5. Moving one column to the right, determine if there are higher-order modifiers from the components of Body, Space, Shape, and/or Effort that are essential to the action. Indicate adjacent to each action stroke if those modifiers are essential, useful, or salient to the action. (See the previously given list of symbols that modify action strokes in these categories.)

  6. How many phrases are there? In other words, which actions belong together to create a phrase? In the next column to the right, use a phrasing bow (see symbols from figure 6.1 in chapter 6) to link those actions.

7. Where is the emphasis, loading, or accent in each phrase? Place an accent mark on the phrasing bow to indicate where the emphasis is or use a phrasing bow that shows even, or increasing, or diminishing. (See symbols from figure 6.1 in chapter 6.)

8. Finally, move one column farther to the right and determine the overall theme of the whole sequence (Inner/Outer, Exertion/Recuperation, Stability/Mobility, Function/Expression, or Self/Other). You may want to move the sequence again to determine this. Draw a theme bracket (see figure 10.9) around the whole motif and use thematic symbols to indicate what you determine. (See figure 3.3 in chapter 3 and the "Exploring the Themes" sections of each chapter in part II.)

9. Move your own motif. Does it capture the essence of your movement sequence? Is there anything you want to add/delete/emphasize? Now that you have done that and created a rich and layered motif of your sequence, what does it mean, and how did it help you to work through this process?

👫 If working in a group:

10. Trade motifs with another student. Move each other's motif (without seeing the movement sequence that it was generated from). Did what you observe the other mover doing capture the essence of the movement sequence that you were trying to motif?

11. Share original movement sequences with each other to determine if the motif needs any adjustment or modifications. What is dominant? What is supportive? What is central to the expression? What modifies the expression? What information is necessary to impart the meaning of your sequence?

How does identifying organization within your movement through motif help you make meaning from it? How many meanings can you associate with your single movement phrase? How did sharing your movement through this medium help you learn new things about the abstract pattern you identified?

- **Interpreting motifs:** In this exercise, you will create movement expression from motif and notice how the motif from which you are creating movement may influence, change, or clarify your movement expression.

  ◦ Return to figure 10.4 and begin with the constellation motif. Create a movement sequence that has all the ingredients put forth in that example.

- ◦ Identify what parts of your movement are fulfilling each symbol in the constellation.

- ◦ Now go to the horizontal motif and order your movement as it is written. Does this change your movement expression and/or intention?

- ◦ Finally, go to the vertical motif. Can you now refine your movement sequence to reflect the relative duration outlined along with the modifications, phrasing, and thematic information? How does this change your movement sequence?

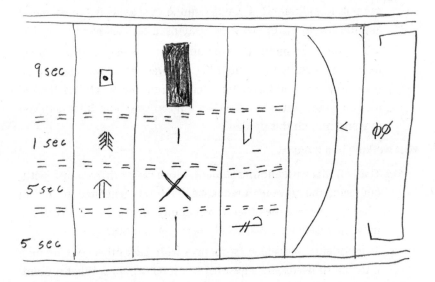

Figure 10.14
Using a CPMMPT staff to notate sequences A and B. This motif is created from sequences A and B in figure 10.6 using the stricter structure suggested in box 10.5, rather than the traditional style shown in figures 10.7 and 10.8. Here, the motif is hand-drawn, as we imagine it would appear in a designer's notebook. This reformatting shows how this structure can flexibly capture the distinct human performances while prioritizing the layers of the movement for easier translation to a foreign body. For example, note how the first action (modified by arc-like directional shape change) does not specify a body part, capturing the different choices by the movers in sequences A (where Amy uses her arm) and B (where Cat uses her head). Moreover, by moving the shape and effort symbols to the Modification column, this idea can be interpreted for (or, in some sense, compiled on) simple platforms (such as the Sphero Mini) as simply any action of the specified duration.

Notice how changing the motif you were moving from may have created different movement sequences. Are there differences in the meaning of your three different sequences based on what motif you were moving from? What would the ultimate "title" be to explain the meaning of each sequence?

- **Designing machine motion:** The motif used in figures 10.7 and 10.8 is redrawn here, in figure 10.14, using the stricter convention introduced in section 10.4. The new motif contains the same information in a more prioritized format that can be better used to translate across bodies, especially with simpler devices like the Sphero Mini. Use this motif to guide your own creation of motion on an artificial body in your environment (either a roboticized body or an inert body that you can puppet with your own body). What does your design capture about the original sequences? What new ideas are introduced in this new embodiment? How did moving your own body help accomplish this task?

Chapter Summary

What does it mean for two distinct bodies to "do the same thing"? This question that we posed at the beginning of this chapter has a complicated answer. On the one hand, given the theories of phenomenology, experience, and perception, we see movement as inherently ephemeral, unique to a given moment in time as well as to the body (or bodies) that produced and perceived it. On the other hand, this chapter has aimed to show how physical practice, training, expertise, and process-oriented observation can create a consistent answer to that question: when we establish a notion of abstraction, we can find similarity between moving bodies. This establishment of the appropriate abstractions for imitation is an essential feature for machines with artificial embodiment. In a fine-grained, objective way, we may use the quantitative abstraction of forgiving microns of machining precision error that define the difference between two identically manufactured and programmed robots in order to declare them to be "the same"; in a coarse-grained, subjective way, we may use a broader abstraction, rooted in movement studies, to establish similarity across distinct bodies and distinct bodily actions on the same body. In doing so, we are beginning to figure out the substance of movement—just as monks in the Middle Ages were figuring out the substance of music—in order to notate it.

Conclusion: Understanding Movement

In this book, we have endeavored to situate movement studies as a crucial part of the landscape of technology research and design. While it is clear that we cannot yet offer a complete structural organization for movement (such as exists in the more fully developed field of music, particularly with regard to notation), we have offered insight into the possibilities that exist for moving in this direction. We have postulated the idea that movement is meaningful to us because of our own experience of our own movement. The idea of the "body as basis" for knowing the world has guided our presentation of material throughout the entire book. Hence, we have offered embodied exercises that make this approach accessible to engineers, designers, and researchers working to produce and refine movement for artificial agents, interpret and predict movement of living organisms, or create and implement harmonious interfaces between the two.

Our goal is to understand how to support a new generation of devices that couple artificial intelligence and artificial embodiment. In this regard, we note with awe that wildly different bodies can be perceived to be doing the same thing. This observation suggests that our goal is possible. To that end, we have suggested using somatic strategies and choreographic technologies to incorporate our own experience of movement in the environment as introduced in part I. This noticing is supported by an extensive taxonomy, introduced in part II, which forms notational abstractions for describing movement. In part III, we outlined case studies of analysis and a notation system—a form of symbolic representation of movement—that supports the ability to transfer and compare movement across different bodies.

Creating Harmony between Humans and Machines

In this book, "movement" has been defined as "perceived change." In highlighting the indivisibility of our perceptions at the outset, we recognize that meaning-making and our experiences, perceptions, and interpretations of movement are ultimately context-dependent. Our lived experience, the site in which the movement we are simultaneously experiencing and perceiving, comes to have meaning only in relationship with the context in which it occurs. Context forms our ability to resolve patterns: an idea comes into perceptual focus when we see its opposite. In other words, meaning is found through resolving paradox (an idea reflected in the topology of a lemniscate).

Perhaps one of the most significant dualities that we understand as important to making meaning with machines is Function/Expression. It lies in recognizing that functional movement is in fact expressive, and expressive movement is in fact functional—that we come to a place of grappling with the full expanse of embodiment. This is an ongoing challenge for any designer: to continue broadening the swath of colors, patterns, and textures available in our design palette. In the context of human perception of movement, the fields related to movement studies, like dance, aim to establish that broad design space.

The summative effect of all the components of the BESST System supports the harmony of movers in their environment, involving both their *kinesphere* and their *dynamosphere*. When we see many people (of the same culture and time) in a shared environment, clear overlaps in their expression within their own kinespheres and dynamospheres occur. That is, the combined interactions of movers in shared spaces creates a dynamic quality inherent to, or associated with, certain spaces. For example, a busy, bustling city intersection feels different from a quiet, rambling creek in the woods. As we build new environments that contain machines moving and monitoring within human-facing spaces, machines will modify and shape the dynamosphere of these spaces (LaViers, 2019c).

The goal of designers in these spaces is the same as the goal of many designers: to create balance and harmony. When we work to increase our movement palette, we create the possibility for novel approaches (and solutions) to emerge. The descriptive taxonomy offered in part II allows us to move away from interpretation and into clearer observation, establishing a more effective design tool for human-facing machine interaction.

Identifying our own habits, preferences, and biases is a key part of this design tool. We are pattern makers and pattern perceivers, and the patterns we make are the patterns we perceive. If we can begin to unpack and understand what those patterns are, the solution space opens wide. Once again, the lemniscate becomes a powerful image of the idea of harmony and balance, capturing the relationship between ourselves and the way that we move, as well as ourselves and the responsive machines in our environment.

Remaining Questions

To this end, here are some of our remaining questions. We hope that suggesting some directions for future work, through the form of some of our own ponderings, helps seed new and ongoing work in creating expressive robots, computer vision for human motion, and other technology design problems that intersect the richness of human movement:

- **Understanding context:** Accounting for temporal, environmental, situational, cultural, and other kinds of context is critical to creating machines that interface with human movement effectively. One way to see the importance of temporal context is to consider repetition. If a movement is perceived to be repeated exactly the same way over and over again, a sense of monotony and inorganic action is created. It is easy to imagine that on the 100th repetition, the action will not be perceived the same way as it was on the first. Any senses of dynamic interest, texture, spatial intent, or mood (to name a few) that may have been present in the first instance may have waned by the 100th. Temporal context is also what creates a sense of phrasing in movement (or the idea that this movement is one, two, or three distinct things). These ideas are important for attempts at simulating effort qualities on artificial bodies. What may be observed as a "flicking" action when encountered alone, may be difficult to discern as a salient moment when accompanied by other light, indirect, and/or sudden actions. Modeling environmental and situational context is also crucial for machine designers. As prior work has shown, movement designed by human animators with one intended label may be labeled entirely differently in a new context (Heimerdinger & LaViers, 2019), although some papers support the idea that certain labels hold up better across context than others (Lambert et al., 2019; Raindel et al.,

2021). This challenges models like those commonly advanced in robot-ics (e.g., see LaViers & Egerstedt, 2012), which suggest that a fixed map-ping between position, velocity, and acceleration of the robot body can generate a sense of effort quality in motion. Which kinds of movements, styles, annotations, or other labels are more context-invariant than oth-ers? How should context (e.g., the situation, environment, narrative, time, place, culture, temporal quality, and spatial quality) be modeled and incorporated into predictive algorithms? How does context interact with higher-order ideas, like those in Effort and Shape, versus lower-level features in Body, Space, and Time?

• **Motif and notation:** Notational abstractions can serve to support transfer of movement across widely differing platforms, but as we discussed in chapter 10, the act of notating is also an act of choosing, subject to the same bias creeping into what is written down. While the current system of motif that we suggest in this book is a tool that can help us translate perceived movement phenomena to other bodies, it is important to rec-ognize that it is also limited—both by the notator and the usefulness of the abstraction. Paying attention to the bias created by our own bodies and sharing our observations with others can be the beginning of resolv-ing this problem. For example, in notating the Sphero Mini discussed in chapter 10, we saw four circles of movement as the salient overall pattern, while the code that generated the movement commanded six distinct cir-cular floor pattern initiations. How do we predict the difference between the code itself and the perceived movement? What sort of staff and sym-bols better capture the movement event? How will different observers notate differently? How will the environment influence the notation?

• **Expression through breath and the core:** The affinities between effort and shape quality highlight the importance of breath and a deform-able spine in human expression. Yet machines rarely include an artifi-cial embodiment of this feature: robots tend to have fixed, immovable core elements and motion models of humans typically create only a few degrees of freedom in this regime. This means that robots cannot express through this postural channel—which, when coupled with limbs, may be more expressive than facial features (Aviezer et al., 2012)—and can-not sense such action in their human counterparts. The complexity of the deformable human core has been modeled using tens of thousands of parameters (Zordan et al., 2004), which should inform the design of machines. How many features does the human core have? How do we measure and model the importance of very, very small movements in animals (e.g., those in the spine)? When, and through what senses, is breath perceived by a human observer?

• **Representation of unique bodies and marginalized demographics:** How we approach design also includes the challenge of inclusion and the representation of marginalized communities; as we have seen, movement is especially personal to the body that is experiencing it. Thus, in finding meaning in motion (either producing it in a robot or interpreting it from data of a human mover), models must contextualize the specific features of the body that is moving. Moreover, experiences from bodies of different demographics than our own (e.g., as described by culture, sex, gender, race, age, and ability) are essential to accommodate in design. By examining our own bias, it may be possible to begin to understand what has been left out—and thus when and where a design may fail. By not attending to the tail ends of a statistical spectrum, we continue to underrepresent those who have been marginalized, and we often marginalize these groups even further in the process. For example, commercial computer vision models have already misidentified certain demographics (Raji et al., 2020), which has significant potential impact on surveillance and law enforcement. Therefore, in designing artificial embodiment, the significance of context and understanding of the experience of all bodies are important to recognize and keep in the foreground of the solution space. How much experimental bandwidth needs to attend to edge cases, minority experiences, and unique circumstances to create tools that work for a wide enough population to be feasible—and fair—to use? What can be done to gain more of these perspectives in the development and validation processes of new technology?

• **Resolving methods and sites for knowledge development:** One of the greatest challenges of this book has been navigating between fields that use distinct methods for transmission of knowledge. The domain of engineering favors objective analysis, quantitative methods, and written texts, while the domain of movement studies favors subjective analysis, qualitative methods, and embodied practice. In writing this book, we have worked to bring in references from fields outside our own core expertise, such as phenomenology, to justify our approach (e.g., the inclusion of embodied exercises as part of the core material). Going forward, the growing space of expressive robotics needs texts and practices that establish it as a site for knowledge development. This suggests a need to increase the writing of practices in movement studies in a way that is accessible to broad application, as well as a need to grapple more with the role that our own embodiment plays in the design of machines. For example, the SlothBot, which was deployed in a public exhibit in the Atlanta Botanical Gardens, has been discussed extensively in terms of the movement mechanism and energy conservation systems on the device, where objective measures of speed and energy capture the advantages of the design. But the machine also contains large, round eyes and other theatrical cues based on the designers' own embodiment and experience of other expressive bodies (the device bears significant resemblance to some popular cartoons) that have not undergone the same rigorous analysis with subjective descriptions of design choices (Notomista et al., 2019). How should the arts, with its focus on subjectivity and quality, be included inside domains like science and engineering, with their focus on objectivity and quantity? How should the experience in the studio or lab be presented in academic journal papers and other archives?

• **Measuring and modeling expressive capacity:** The BESST System has been developed to help people express themselves better through movement: training with the Basic Six enhances muscular efficiency; moving through the different geometric forms enumerated by the scales encourages bodily balance and range of motion; using phrasing creates more clarity in the structure of their movement; adding shape qualities to their expression of spatial pulls enhances the sense of "up," "left," "right," and so on; and developing a robust recall for images and ideas can help them enact specific qualities of movement. In theory, we can do the same for robots. But how can we measure and model this capacity? The static, kinematic measure proposed by LaViers (2019a) is one step toward this, but it does not relate the measure to the parameters given in the BESST System, nor does it account for dynamic changes of a body in time. Which elements of the system can be executed by a given robot? It is easy to measure the force capacity of a given actuator, but how does this measure translate to expression?

A Crucial Body of Expertise

Throughout this book, we have mostly avoided using the term "tacit knowledge." Meaning "understood or implied without being stated," it is often used to describe physical knowledge or intuition. This idea complicates much of our perspective on movement studies. On the one hand, we have emphasized that not everything can be written; some things need to be experienced. To this end, we have presented physical practice as a form of research, even entreating you, our reader, to engage in such activity in order to come to a full understanding of the material. On the other hand, we have worked to exhaustively catalog as much of what we know about movement as we can in these limited pages, painstakingly trying to write, formalize, and codify expertise in movement as a perceived phenomenon. However, as we have seen, notation falls short of capturing abstract representation of movement expression.

Describing embodiment as "tacit" can feel like taking a shortcut or a way to avoid doing the work of explicating knowledge, which has the double-edged effect of pushing this type of investigation out of the academy. At the same time, the notion that *not everything can be written down* feels right to us and guides our method of giving workshops in big, open rooms, wearing comfortable clothes, and practicing movement in our bodies. These points of view are not in conflict. Indeed, there are many physical phenomena that we cannot explain, cannot write, or cannot model with equations, but such phenomena are not written off as impossible to explain; instead, we attempt to understand them. There may be no better place to try to understand human movement than alongside machines, instruments of movement, just as writers, painters, and musicians came to an understanding of their craft through the implements and tools of their pursuits. That is, not everything can be written down *yet*.

To explicate our embodied experience, we must give attention to the methods shared in this book and add to the emerging intersection between robotics and movement studies. That is, as movers, we need to take the time to try to write down the results (externalizing our experience with objectivity) and, in parallel as scholars, we need to give legitimacy to research conducted through practice (allowing internal perspectives to be shared with subjectivity). Such a charge amounts to using structured approaches that are more common to engineering and science inside movement studies and the arts and personal perspectives more common to the arts and humanities inside machine design. Our inability to notate and represent movement's expressive dimensions points to an urgent dearth of knowledge in both fields, creating a mystery around movement parameterization and analysis.

Where do we find the clues to this mystery? In every movement practice imaginable: in the hands of expert weavers in Peru, in the hips of college wrestlers, in the intricately shaped hands of bharatanatyam dancers, in the wrists of pastry chefs, in the feet of Paralympians, in the fingers of computer programmers, and in the many, many more physical practices that make up human achievement. These communities—like choreographers—are constantly developing new strategies and technologies to push their respective activities to new heights.

These communities can struggle to be represented in academia. Troubling trends include the elimination of performing arts programs, the fewer and fewer sports represented in college athletics, and higher and higher burdens on student time and performance, making embodied activities feel in competition with other academic studies. Thus, not only do we need collaboration across disciplines *within* academia, we need collaboration *outside* academia with industrial partners, practitioners, athletes, and community members.

We predict that the technological fields dealing with embodiment, like robotics, artificial intelligence, and kinesiology, will not succeed in integrating their products in human-facing scenarios without these external resources. These fields face too many somatic and choreographic challenges that—in our experience—cannot be solved without bodily investigation. Thus, we dream of pursuits in movement flourishing alongside scholarship, empirical investigation, and personal meaning-making.

Practitioners of various somatic and choreographic practices contain clues to, and understand pieces of, the symbolic, systematic understanding of human movement. We have focused on the Laban/Bartenieff tradition in this book, but we want to show that there are methods beyond it that help organize and describe movement. We point to open problems through this review and broaden our consideration to include sports, physical therapy, social dance, and other ways of "knowing" as crucial forms of human knowledge.

When it comes to understanding human movement, we are in the times before cuneiform (early language symbology), before the treble/bass clefs (early music symbology). In movement, we have developed many functional measures, methods, and models, particularly in engineering. But as we have seen, function is only half the picture. In writing and music, it was through studying expression that the systems that underpin language and sound were developed; thus, the legitimacy, potency, and value of physical and expressive practices have never been more important.

Outro: Returning to Embodied Perspectives

Finally, we return to embodied perspectives like those that opened the book in the prelude. This time, the perspectives are enriched with concepts from the taxonomy introduced here, and as before, they represent the critical aspects of the site of work for this domain: embodied, personal, and contextual. Cat writes of her experience in the Certification in Movement Analysis (CMA) certification program with Ellen Goldman at the Laban/Bartenieff Institute of Movement Studies (LIMS) in New York in the fall of 1983, and Amy writes about a recent experience in a yoga class with Leo Eisenstein at The Bridge Hot Yoga, in Narberth, Pennsylvania, in the winter of 2021.

The first challenge in the certification program includes being assessed on the execution of the diagonal scale; for clarity in space, then with effort, and finally a "personal" version that is expressive and unique to the mover. I felt fairly confident about the spatial execution; I had memorized the sequence and aligned it in my mind to the corners of the room—essentially imagining the room as my "cube" and the corners as my spatial "points." I demonstrated the sequence and was met with some surprise and quizzical reactions from Ellen. She told me that I clearly knew the sequence but was simply indicating the places in the room rather than revealing the diagonal space. She encouraged me to feel the space not only moving through my body, but beyond my body, thus revealing both the spatial pulls and my engagement and relationship to them. I brought my full sensing, thinking, feeling self to the task, and for the first time, I experienced that space was an active partner with me and my movement expression. By using the support of shape qualities, rotation, and active weight shift and support, I was able to dance with the space and reveal not only the complex, three-dimensional pulls that I was expressing from my body to the space, but the pulls of the space and the effects of that in my body. It was the first time that I truly understood my relationship to the environment and the nuanced ability that I had to "partner" from inner to outer, from body to space, to truly reveal the dynamics and pulls of complex, three-dimensional movement. This process began my journey toward understanding and experiencing the power, beauty, and fullness of dancing with the environment—and the myriad ways that the environment shaped me and I shaped the environment. I began to understand that space is, in and of itself, meaningful.

Cat

It's Saturday, and I am at yoga. Usually I attend on Sunday, so I am adjusting to Leo's dialogue. He is chattier than Chalise, my usual instructor, and at times more opinionated on the expression of each posture. I have to remember to foreground myself, negotiating the Self/Other duality: I am here for me, not to please the instructor, fawning over each offered correction to display my earnestness, even though my striving often reveals something new about myself. Thus, I start with my feet, kneading them to awaken the haptic and proprioceptive channels within, finding flexion, extension, and rotation in as many of these distal joints as I can. And, when class begins, I try on his dialogue, his movement instructions, spoken aloud to the whole class, as a source for new ideas to integrate into my practice. This class employs a series that is always the same twenty-six postures—a series I've been practicing for over ten years. Yet, across ebbs and flows in my attention, fitness, and flexibility, I discover something new every time I practice. Surely this alone speaks volumes about the enormity of the bandwidth required to describe the possible human experience of movement. The first transformative insight comes from Leo's directions in the fourth posture: eagle. Like all the postures, I go through phases with eagle inspired both from concepts that originate within me and concepts shared by my teacher, each folding at the nexus of Inner/Outer, creating the thing that I do in class, the movement that I enact in the environment of heat, sweat, and instructions.

The pretzel-like posture involves twisting both arms and legs into tight spirals while bending low, as if sitting in a chair, with a single foot placed flat on the ground. Initially, I approached moving into the posture almost like a ballet battement, holding my leg bent (in attitude) but keeping it extended in far-reach space as I lifted it off the ground, before curling it inward to wrap around my standing leg to touch the other leg in several places. Then, a teacher in Atlanta told me to relax the lower leg below the knee, reducing the entry to an action in mid-reach space, giving more relaxation to my hip socket and allowing a deeper hip flexion (as in the thigh lift of the Basic Six). Now, Leo is inviting me to do the action as one long, integrated movement instead of two or three. That is, he suggests that I change the phrasing of the movement. It's an interesting suggestion that I hadn't considered before, so I try it. The posture feels completely different. What was a piecewise string of weight-sharing between my legs is now an integrated twisting, allowing me to access place low in a more direct and connected way. I feel as though I'm less like a tower of off-centered blocks and more like a system of elegant roots growing into the ground. On another day, returning to chunking the act into three smaller motions may give me greater access to finding the twisted weight sharing between joints that opens the sockets and leaves me feeling taller. But today, this new option is a revelation.

Amy

Appendix A: Symbols in the BESST System

Components

Body	8
Space	☐
Time	|
Shape	//
Effort	/

Themes

Inner/Outer	⊗Q
Inner	⊗⊖
Outer	⊖Q
Stability/Mobility	⏀⊘
Stability	⏀⊖
Mobility	⊖⊘

Function/Expression ⬭⬭

Function ⬭⬭

Expression ⬭⬭

Exertion/Recuperation ⬭⬭

Exertion ⬭⬭

Recuperation ⬭⬭

Self/Other ⬭⬭

Self ⬭⬭

Other ⬭⬭

BODY

Breath

Breath ⊙

Inhale ⊙

Exhale ⊙

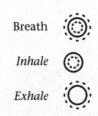

Features of Expressive Bodies

Axis of length ⬥

Upper ⬥

Lower ⬥

Midline ⬥

Left ⬥

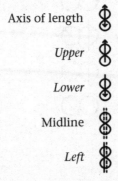

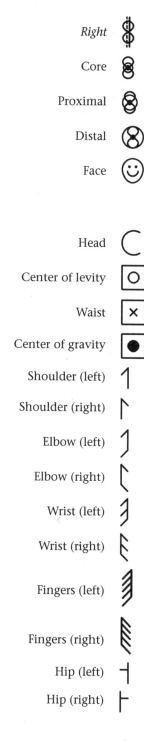

Right

Core

Proximal

Distal

Face

Body Parts

Head

Center of levity

Waist

Center of gravity

Shoulder (left)

Shoulder (right)

Elbow (left)

Elbow (right)

Wrist (left)

Wrist (right)

Fingers (left)

Fingers (right)

Hip (left)

Hip (right)

Knee (left)   ⊣

Knee (right)   ⊢

Ankle (left)   ⊒

Ankle (right)   ⊨

Toes (left)   ⫸

Toes (right)   ⫶

Patterns of Body Organization

Radial symmetry

Spinal

Core/distal

Head/tail

Upper/lower

Right/left

Cross-lateral

Flow-/Weight-Sensing

Weight-sensing

Flow-sensing

Basic Body Actions

Change of support

Change of support in series

Axial Movements

Posture ⚮

Gesture ⚮

Condense ✕

Expand Ͷ

Rotate ᛗ

Hold ○

Focus ⊗

Vocalize ⌄

Touch ⌄

Locomotor Movements

Travel Ⅰ

Roll Ⅰ

Slide Ⅰ

Walk ╌Ⅰ╌

Run ╌Ⅰ╌

Jump: any (᛫)

1-to-1: same ( )( )

1-to-1: other ( )( )

1-to-2 (᛫᛫)

2-to-1 (᛫)

2-to-2 (᛫᛫)

SPACE

Zones

High

Low

Right

Left

Front

Back

Planes

Vertical plane |V|

Sagittal plane |S|

Horizontal plane |H|

Reach Space

Near

Mid

Far

Pathways

Central

Peripheral

Transverse

Spatial Direction

Middle Plane

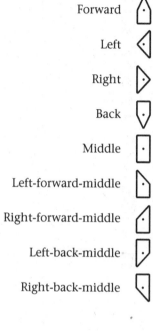

Forward

Left

Right

Back

Middle

Left-forward-middle

Right-forward-middle

Left-back-middle

Right-back-middle

Low Plane

Forward-low

Left-low

Right-low

Back-low

Place-low

Left-forward-low

Right-forward-low

Left-back-low

Right-back-low

High Plane

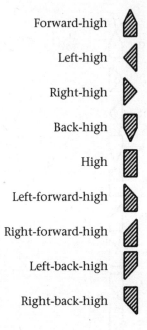

Forward-high

Left-high

Right-high

Back-high

High

Left-forward-high

Right-forward-high

Left-back-high

Right-back-high

TIME

Phrasing

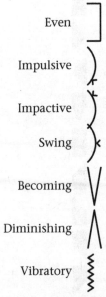

Even

Impulsive

Impactive

Swing

Becoming

Diminishing

Vibratory

SHAPE

Still Shape Forms

Pin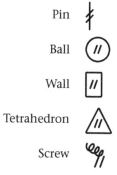

Ball

Wall

Tetrahedron

Screw

Primary Shape Patterns

Concave) (

Convex ( )

Gather JL

Scatter TT

Modes of Shape Change

Shape flow ─//─

Directional ─//→

Arc-like ─//⌐

Spoke-like ─//↗

Shaping ─//⌣

Inner shaping

Shape Qualities

Rising	
Spreading	
Enclosing	
Retreating	
Advancing	
Sinking	

Combinations of Two Qualities

Rising and spreading	
Rising and enclosing	
Rising and advancing	
Rising and retreating	
Sinking and spreading	
Sinking and enclosing	
Sinking and advancing	
Sinking and retreating	
Spreading and advancing	
Spreading and retreating	
Enclosing and advancing	
Enclosing and retreating	

Combinations of Three Qualities

Rising, spreading, and advancing

Rising, spreading, and retreating

Rising, enclosing, and advancing

Rising, enclosing, and retreating

Sinking, spreading, and advancing

Sinking, spreading, and retreating

Sinking, enclosing, and advancing

Sinking, enclosing, and retreating

EFFORT

Factors

Weight

Light

Strong

Flow

Free

Bound

Space

Indirect

Direct

Time _/_

    Sustained _/

    Sudden /_

States

Mobile ⇤

  Bound and sudden ⇙

  Bound and sustained _⇙

  Free and sudden ⟋_

  Free and sustained ⟋

Remote ⊥

  Bound and direct ⊏

  Bound and indirect ⌋

  Free and direct ⌐

  Free and indirect ⌡

Dream +

  Strong and bound ⌐

  Strong and free ⊤

  Light and bound L

  Light and free ↓

Stable ⊢

  Strong and direct ⌠

  Strong and indirect ⌡

  Light and direct ⊢

  Light and indirect ⌡

Rhythm

Strong and sudden

Strong and sustained

Light and sudden

Light and sustained

Awake

Direct and sudden

Direct and sustained

Indirect and sudden

Indirect and sustained

Drives

Action drive

Action drive float

Action drive punch

Action drive glide

Action drive slash

Action drive dab

Action drive wring

Action drive flick

Action drive press

Passion drive

Passion drive float

Passion drive punch

Passion drive glide

Passion drive slash

Passion drive dab

Passion drive wring

Passion drive flick

Passion drive press

Spell drive

Spell drive float

Spell drive punch

Spell drive glide

Spell drive slash

Spell drive dab

Spell drive wring

Spell drive flick

Spell drive press

Vision drive

Vision drive float

Vision drive punch

Vision drive glide

Vision drive slash

Vision drive dab

Vision drive wring

Vision drive flick

Vision drive press

Appendix B: Movement Scales

The movement scales use the crystalline forms introduced in chapter 5 (octahedron, cube, and icosahedron)[1] and their infrastructures to create sequences of movement that progress in different arrangements through those forms. In moving a series that is designed to contain spatial pulls of only one, two, or three dimensions, we can explore the idea of what feels comfortable and common versus uncomfortable and rare. The answer to this conundrum is probably both personal and specific to common human anatomy. It is thought that it is more common for human motion to comprise transitions between two unequal spatial pulls. Moreover, the particular transition that moves from one plane (say, the vertical) to another plane (say, the sagittal), while requiring transit through the third one (in this case, the horizontal) requires a bodily accommodation that is thought to be a key signature of the three-dimensionality of human movement. For example, right-high to forward-low requires the mover to pass through the horizontal plane on the way from the vertical plane to the sagittal plane. These special movements necessarily use transverse pathways through the kinesphere and are termed **transversals.**

- Octahedron (one spatial pull at each step of the scale; see figure B.1):
  - The **dimensional scale** can be performed by using central pathways only (going through place middle before expressing the next pull) or alternating central and peripheral pathways (going through the center of the body when expressing each dimension and then along the edge of the octahedron when changing to another dimension).
    - When the dimensional scale is executed with all central pathways, there is a "poking out and in" character to the expression. The movement goes away from the body and then back into the neutral

place middle before moving to another dimension. In/out, in/out is the rhythmic pattern created by experiencing the scale in this way. There is a unipolar experience of the dimensions, which can also produce a sense of more parts to the whole scale.

- When the dimensional scale is executed with alternating central and peripheral pathways, the expression of a full dimensional pull is experienced as bipolar because the central pathways connect both pulls in the dimension and the peripheral pathways link one dimension to the next by moving along the edge of the octahedron. The rhythm has more of an in/out/edge feeling, which can also produce a sense of fewer parts to the whole scale.

○ The **defense scale** is performed using peripheral pathways only, based around zones of the body, as loosely seen in fencing and attack/defense motions.

- This scale consists of all peripheral pathways that move along the edges of the octahedron and is more outer in its expression, as the expression of spatial pulls never passes through the neutral place middle. The dimensions are expressed by passing through another dimensional pull, creating a clear delineation of the octahedron. As such, there is a sense of boundary, of creating a container of the space for the expression to occur. The rhythm is more swooping, and the scale feels as if it only has three parts.

• Cube (three equal spatial pulls at each step of the scale; see figure B.1):

○ The **diagonal scale** alternates using central and peripheral pathways. The full expression of each diagonal (two equal and opposite three-dimensional pulls) is revealed as a central pathway, and as the mover changes to the next diagonal, the pathway is peripheral along a face of the cube itself. This scale focuses on the rarefied expression of three-dimensional pulls in equal expressions of the vertical, sagittal, and horizontal simultaneously.

- As each diagonal is expressed in its entirety, through a central pathway, and the transition to the next diagonal is expressed as a peripheral pathway, along a face of the cube, there is an feeling of in/out/face in the rhythm that creates a mobile expression, which enhances the sense of the volume of space around the body. For example, going from right-side-high to left-back-low (central pathway) and from left-back-low to left-forward-high (peripheral pathway) takes the mover into the full left side zone of space.

- The sequence is ordered as right-forward-high to left-back-low (central), left-back-low to left-forward-high (peripheral along the left face of the cube), left-forward-high to right-back-low (central), right-back-low to left-back-high (peripheral along the back face of the cube), left-back-high to right-forward-low (central), right-forward-low to right-back-high (peripheral along the right face of the cube), right-back-high to left-forward-low (central), and left-forward-low to right-forward-high (peripheral along the front face of the cube), to end where it began.

The pattern of gradual and abrupt change within these sequences creates a sense of rhythm to each scale. Within the prescribed order of relationship of the various spatial pulls in each scale sequence, different qualities of expression emerge. The dimensions are more stable, while the diagonals are more mobile; hence, the octahedral scales often feel precise, measured, and more linear, while the diagonal scale in the cube is highly mobile (pathways with three equal pulls abruptly changing, alternating with pathways in which two spatial pulls are changing) and is often characterized as being about "flying and falling." This is also attributed to the fact that because our human form is not dimensionally equal (we are usually more vertical than anything else), the expression of three equal spatial pulls simultaneously is a rarefied and difficult moment to sustain.

- Icosahedron (two unequal spatial pulls at each step of the scale):
  - **Moving through the planes** (9): Cycling the three planes that form the vertices of the icosahedron is one way to practice sequencing spatial pulls. These involve moving from each of the "corners" of the vertical, sagittal, and horizontal planes. These movements can happen by avoiding middle or by moving through middle, bisecting the plane and creating different patterns. See figure B.2.
    - *Vertical plane*
      - Cycle (with peripheral pathways)
      - Bisect (starting central pathway right-high to left-low, and then peripheral pathway left-low to right-low)
      - Bisect (starting peripheral pathway right-high to right-low, and then central pathway right-low to left-high)

- *Sagittal plane*
  - Cycle (with peripheral pathways)
  - Bisect (starting central pathway forward-high to back-low, and then peripheral pathway back-low to forward-low)
  - Bisect (starting peripheral pathway forward-high to forward-low, and then central pathway forward-low to back-high)
- *Horizontal plane*
  - Cycle (with peripheral pathways)
  - Bisect (starting central pathway right-forward-middle to left-back-middle, and then peripheral pathway left-back-middle to left-forward-middle)
  - Bisect (starting peripheral pathway right-forward-middle to right-back-middle, and then central pathway right-back-middle to left-forward-middle)

o **Axis scales** (4): These are transverse scales that move from one planal direction to another planal direction in relation to a specific diagonal, creating a sort of "basket" around a specific diagonal. Each pathway is a transversal and can be characterized by being either "flat" (going from the horizontal plane to the vertical plane), "steep" (going from the vertical plane to the sagittal plane), or "suspended" (going from the sagittal plane to the horizontal plane). Thus, the "feeling" or "character" of the pathways begins to be revealed as different from each other. All these scales prescribe an order of vertical to sagittal to horizontal space. See figure B.3.

  - Around the right-forward-high to left-back-low diagonal (traditionally beginning right-high)
  - Around the left-forward-high to right-back-low diagonal (traditionally beginning left-high)
  - Around the left-back-high to right-forward-low diagonal (traditionally beginning right-low)
  - Around the right-back-high to left-forward-low diagonal (traditionally beginning left-low)

o **Girdle scales** (4): These are peripheral scales and move around a specific diagonal with the pulls that are not actual deflections of that diagonal.

Thus, there is a relationship between the axis scale for a specific diagonal (planal deflections of that diagonal) and its girdle scale (planal pulls that are *not* the deflections of that diagonal). These scales also follow a prescribed order of vertical to sagittal to horizontal space. See figure B.3.

- Around the right-forward-high to left-back-low diagonal (traditionally beginning left-high)
- Around the left-forward-high to right-back-low diagonal (traditionally beginning left-low)
- Around the left-back-high to right-forward-low diagonal (traditionally beginning left-low)
- Around the right-back-high to left-forward-low diagonal (traditionally beginning left-high)

○ **Primary scales** (4): These are the peripheral pathways using all twelve icosahedral directions around each diagonal and as such form a sort of chain around a diagonal. These scales are formed by alternating an axis scale direction with a girdle scale direction of each diagonal. These scales follow a prescribed order of vertical to horizontal to sagittal space. See figure B.4.

- Around the right-forward-high to left-back-low diagonal (traditionally beginning back-high)
- Around the left-forward-high to right-back-low diagonal (traditionally beginning back-high)
- Around the left-back-high to right-forward-low diagonal (traditionally beginning back-low)
- Around the right-back-high to left-forward-low diagonal (traditionally beginning back-low)

○ **A and B scales** (2): These are transverse scales that use all twelve icosahedral directions relating to three diagonals. The "missing" diagonal becomes the axis around which these scales rotate. The A and B scales mirror each other. See figure B.4.

- A scale
  - Missing the flick-press diagonal
- B scale
  - Missing the glide-slash diagonal

The prescribed order of these scales follows vertical to sagittal to horizontal space and can further be experienced with different emphasis on how the mover groups the visited spatial pulls. For example, a sequence of two pathways that relate to different diagonals is called a **volute** and produces a motion between three pulls that is more sweeping and rounded. A series of two pathways that are deflections of the same diagonal is called a **steeple** and is experienced as a more pointed, two-part experience of a trio of pulls.

The icosahedral scales each have a different character unto themselves, which is experienced as an aspect of the larger ideas of gradual versus abrupt change, their relationship to the planal sequencing and the notion of the different qualities of various pathways (in this case, peripheral pathways versus transverse pathways), and finally the relationship of each scale to a particular diagonal. Each plane (as well as the cycle of its spatial pulls) creates a different experience. The space of the vertical plane is presentational (performing arts like ballet and public oration foreground this plane through the use of proscenium space); the space of the sagittal plane is more action or decision oriented (locomotion and travel happen primarily in this plane); and the space of the horizontal plane is more communicative (actions like conversational gestures that accompany banter at mealtime are necessarily constrained by the plane of a dinner table).

The A scale rotates around the missing diagonal of right-back-high to left-forward-low, which Laban (1966, p. 158) felt was the most "strong and impulsive" of the four diagonals. So the "most powerful" expression of diagonal space is not present in the scale sequence. In addition, the sequencing of planal organization forms a more "defensive" feeling, which Laban further characterized as "feminine," likening it to the minor scales of music. In contrast, the B scale rotates around the missing diagonal of left-forward-high to right-back-low, which is characterized as "gentle." The planal sequencing of the B scale forms a more "attacking or aggressive" feeling, which Laban characterized as "masculine" and likened to the major scales of music. Whether or not these characterizations resonate with each mover, it is apparent that the quality of expression and experience is different between these two scales.

The axis scales run up and down a particular diagonal; Laban characterized them as having a more "unconscious" or "automatic" aspect of experience that he likened to falling asleep while sitting and then jerking back to awakeness, or the stumbling of a drunk person lurching their way to a place of rest. In contrast, the girdle scales, with their circular peripheral pathways, promote a feeling of "alertness" and clarity of consciousness. The sense of defining a container emerges in distinction to the sense of moving contents that is more prevalent in the axis scales. There is a dichotomy between the more inner experience of the axis scales and the more outer experience of the girdle scales.

The primary scales combine the axis and girdle scales and as such have a more serpentine, meandering quality of short pathways that alternates between a more inner, unconscious experience and a more outer or alert consciousness. There is an experience of modulation, a balance of alertness and unconsciousness.

Embodied Exercises

- **Exploring the A and B scale in relationship to each other:** In this exercise, you will move through the first part of the A scale and the first part of the B scale from differing starting points to experience a change in spatial expression.
  - Leading with the right side of the body, move the following sequence from the A scale:
    - Right-side-high to back-low to left-forward-middle to right-side-low to back-high. (This experience lends itself to finding the volute phrasing of the scale.)
    - Try the same sequence, but start at back-low and finish at right-forward-middle. Does your experience of the space and the story it tells change? How? (This experience lends itself to finding the steeple phrasing of the scale.)
  - Leading with the right side of the body, move the following sequence from the B scale:
    - Left-side-high to forward-low to right-back-middle to left-side-low to forward-high (volute phrasing).
    - Try the same sequence but start at forward-low and finish at left-back-middle (steeple phrasing). Does your experience of the space and the story it tells change? How?

For these experiences, try moving the space with the engagement of the limbs, and then just by moving your core with no limb involvement. What changes in your experience? Once you have explored all the sequences, make some notes on what you perceived and experienced for each. What are the similarities and/or differences between these four sequences? Do you experience a different expression for each one? What do you perceive as their relationship to each other? To your body? You can refer back to the entirety of each of these scales and notice how the expressive quality of the space changes based on sequence, emphasis, and the way that you engage your body in the space.

- **Explore spatial and temporal phrasing with gradual/abrupt change:** Revisit the A and B scales and explore the phrasing of the entire sequence in the following way:
  - With volute phrasing (for the B scale, right side leading):
    - Move from left-side-high to forward-low to right-back-middle, as one phrase.
    - Move from right-back-middle to left-side-low to forward-high.
    - Move from forward-high to left-back-middle to right-side-low.
    - Move from right-side-low to back-high to left-forward-middle.
    - Move from left-forward-middle to right-side-high to back low.
    - Move from back-low to right-forward-middle, ending where you began at left-side-high.
  - With steeple phrasing (for the B scale, right side leading):
    - Move from right-forward-middle and go to left-side-high, ending in forward-low, moving this as one phrase.
    - Move from forward-low to right-back-middle, ending left-side-low.
    - Move from left-low to forward-high to left-back-middle.
    - Move from left-back-middle to right-side-low to back-high.
    - Move from back-high to left-forward-middle to right-side-high.
    - Move from right-side-high to back-low, ending where you began in right-forward-middle.
  - Do you experience this larger phrase as six movements or more? Does each bullet point feel like one movement or two?

- Volute phrasing is often characterized as rounded, swooping with a triplet rhythm—can you practice the scale to experience this? Steeple phrasing is often characterized as jagged, zigzag, pointed, with a duple rhythm—can you practice the scale to experience this?
- How do these distinct choices in spatial phrasing affect your temporal performance of the scale? Do you perform the bullet points in the volute phrasing more slowly than the bullet points in the steeple? If not, try this. How does this support a sense of gradual change in the volute execution and a sense of abrupt change in the steeple execution?
- Reflect on the rhythmic structure relating to both the space being expressed and the organization in time that is being used.

Dimensional Scale (Central Pathways)

Dimensional Scale (Alternating Central/Peripheral Pathways)

Defense Scale (Peripheral Pathways)

Diagonal Scale (Alternating Central and Peripheral Pathways)

Figure B.1
Dimensional, defense, and diagonal scales (with right side leading) presented as horizontal motifs with a preparatory spatial pull (from where one should begin the scale) provided.

Vertical Plane: V

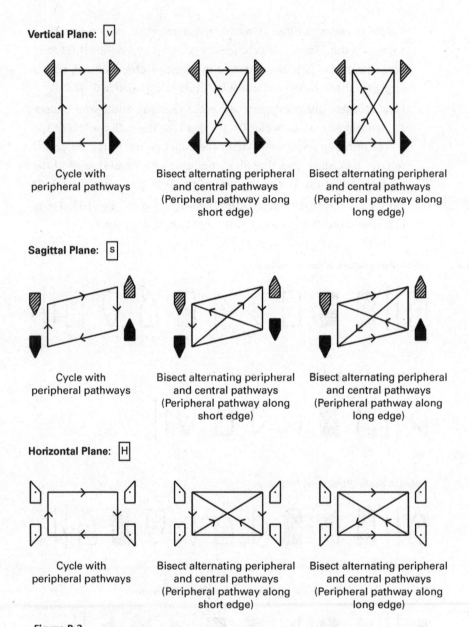

| Cycle with peripheral pathways | Bisect alternating peripheral and central pathways (Peripheral pathway along short edge) | Bisect alternating peripheral and central pathways (Peripheral pathway along long edge) |

Sagittal Plane: S

| Cycle with peripheral pathways | Bisect alternating peripheral and central pathways (Peripheral pathway along short edge) | Bisect alternating peripheral and central pathways (Peripheral pathway along long edge) |

Horizontal Plane: H

| Cycle with peripheral pathways | Bisect alternating peripheral and central pathways (Peripheral pathway along short edge) | Bisect alternating peripheral and central pathways (Peripheral pathway along long edge) |

Figure B.2
Cycling and bisecting the planes (with right side leading) presented with geometric renderings of the planes overlaid with spatial pull symbols.

Axis Scales (Transverse Pathways)

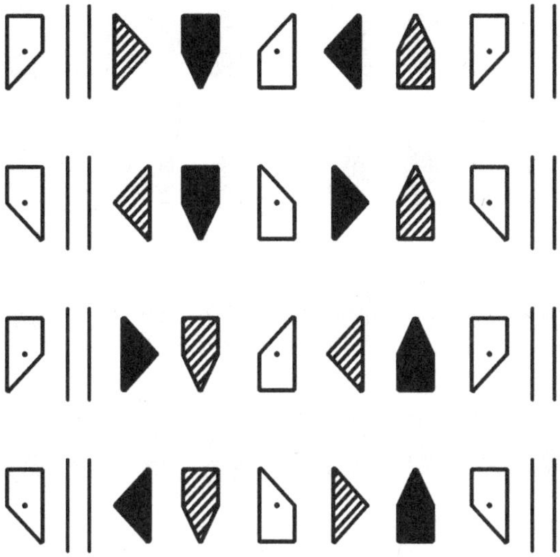

Girdle Scales (Peripheral Pathways)

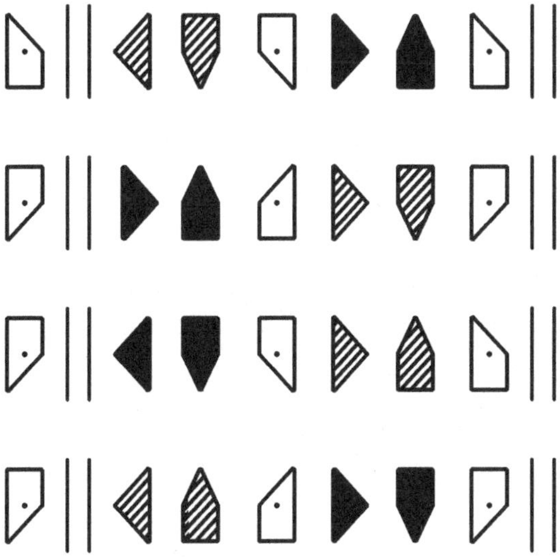

Figure B.3
Axis and girdle scales (with right side leading) presented as horizontal motifs with a
preparatory spatial pull (from where one should begin the scale) provided.

Primary Scales (Peripheral Pathways)

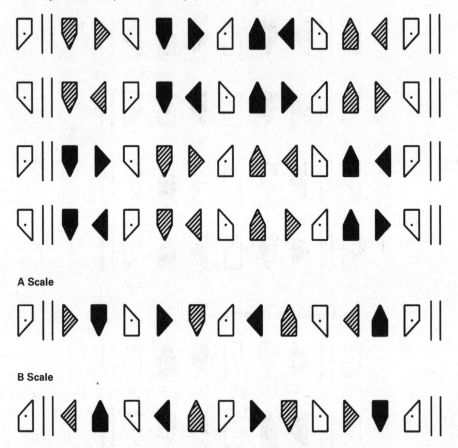

Figure B.4
Primary, A, and B scales (with right side leading) presented as horizontal motifs with
a preparatory spatial pull (from where one should begin the scale) provided.

Appendix C: Effort Configurations

Effort states use combinations of two effort factors, listed in parentheses after each. An image that may help elicit the movement associated with that quality is also listed for each combination within the state. This constitutes an initial "effort bank." Please feel free to write your own images in the margins of this appendix as you find them, because such images are personal, not universal.

Effort transformation drives use flow effort to create variations of the basic effort actions (BEAs), as listed and defined in section 8.1 of chapter 8. Box C.2 lists each, along with the combinations of three effort factors; an image that may help elicit the movement associated with that quality is also listed for each combination within the drive. Each drive contains eight combinations, using the same eight BEA labels. They are distinguished from one another by their drive. For example, an *action drive punch* versus a *passion drive punch*. In the transformation drives, flow effort replaces each factor in action drive in turn: passion drive swaps flow for space, vision drive swaps flow for weight, and spell drive swaps flow for time.

Embodied Exercises

- 📹 **Using familiar images to invoke effort:** Use the bullet points to provide you with images often (but not always) associated with these states in order to explore movement qualities.
  - Move your body in response to each image.
  - Record (or film) your response.
  - Observe your response. Is your intent manifest in the observed motion? How could you edit your performance to improve?

Box C.1
Effort States with Relevant Images

- ○ **Mobile state** (flow and time)
  - Being startled by something and freezing in response (bound/sudden)
  - Pulling out a stretchy, pliable material, like taffy, to a desired length (bound/sustained)
  - Kernels of popping popcorn (free/sudden)
  - Sighing in relief (free/sustained)
- ○ **Stable state** (weight and space)
  - Saying "No" firmly and pointedly, perhaps with an open palm outstretching at the same time (strong/direct)
  - Scrubbing a dirty floor (strong/indirect)
  - Inserting a contact lens into an eye (light/direct)
  - Stroking a kitten's fur (light/indirect)
- ○ **Remote state** (flow and space)
  - Threading a needle (bound/direct)
  - Icing a cake with swirls of frosting (bound/indirect)
  - Blowing out a bunch of birthday candles on a cake (free/direct)
  - Standing in awe on the summit of a mountain, absorbing the grand, 360-degree vista (free/indirect)
- ○ **Rhythm state** (weight and time)
  - Bouncing on a diving board (strong/sudden)
  - Moving a grand piano or heavy piece of furniture (strong/sustained)
  - Gasping with a shiver (light/sudden)
  - Savoring a delicious cookie (light/sustained)

Note how personal and contextual this is: we both resonate with this image today, but if we were coming back from the woods from a long hike with little food, the way that we would savor would probably engage a strong sense of weight.

- ○ **Dream state** (weight and flow)
  - Clenching your fist (strong/bound)
  - Shouting "Hello!" in a large empty canyon (strong/free)
  - Blotting your sweat, avoiding smudging your makeup (light/bound)
  - Blowing a hair off your face (light/free)
- ○ **Awake state** (space and time)

(*continued*)

Box C.1 (continued)

- Looking to someone who has just called your name (direct/sudden)
- Viewing an engaging piece of visual art (direct/sustained)
- Startling in response to a sound and not knowing where it came from (indirect/sudden)
- Scanning a large crowd of people to find a friend (indirect/sustained)

Box C.2
Transformation Drives with Relevant Images

- **Passion drive** (weight, flow, and time)
  - Float (light/free/sustained): filled with rapture
  - Punch (strong/bound/sudden): recoiling from horror
  - Glide (light/bound/sustained): the moment of internal focus before sneezing
  - Slash (strong/free/sudden): a wet dog shaking off all the water from its fur
  - Dab (light/bound/sudden): working to control a hiccup
  - Wring (strong/free/sustained): waking up stretching after a fitful night of bad dreams
  - Flick (light/sudden/free): a fairy delicately sneezing
  - Press (strong/bound/sustained): the aching heartbreak of a martyr
- **Spell drive** (weight, space, and flow)
  - Float (light/free/indirect): a witch spreading her fairy dust over the world
  - Punch (strong/bound/direct): a witch cursing her sworn enemy with an incantation
  - Glide (light/free/direct): turning a page while engrossed in a novel
  - Slash (strong/bound/indirect): clearing your way through an overgrown haunted forest; seeing creepy eyes peering at you from everywhere
  - Dab (light/bound/direct): Dorothy in *The Wizard of Oz* clicking her heels together, chanting "There's no place like home"
  - Wring (strong/free/indirect): a dragon razing an entire community with its fiery breath
  - Flick (light/bound/indirect): reveling in the experience of clearing foggy condensation off a large and delicate windowpane
  - Press (strong/free/direct): a dragon breathing fire at a particular tree

(*continued*)

Box C.2 (continued)

∘ **Vision drive** (flow, time, and space)

- Float (free/sustained/indirect): a landscape architect dreaming up a new design

- Punch (bound/sudden/direct): grasping a falling glass before it shatters

- Glide (free/sustained/direct): the opening scroll of text in *Star Wars*: "a long time ago in a galaxy far, far away"

- Slash (bound/sudden/indirect): gathering strands of sugar out of a cotton candy machine

- Dab (free/sudden/direct): trying to count the number of fireflies you see lighting up

- Wring (bound/sustained/indirect): surveying a plot of land

- Flick (free/sudden/indirect): gathering children scattered across the yard inside to protect them from an oncoming storm

- Press (bound/sustained/direct): staring someone down in a staring contest; locking a gaze with intensity

---

∘ Repeat this cycle to refine your palette of dynamic qualities. As the nuance in your performance grows, so does the nuance in your ability to observe.

- **Understanding your personal relationship to effort:** Create your own personal images to help invoke different configurations of effort. Building such an "effort bank" is an essential part of training for movement analysts.

Appendix D: Rationale for the Time Component

In prior texts on movement studies, Laban Movement Analysis (LMA), and Laban/Bartenieff Movement Studies (LBMS), as well as in movement analysis certification programs, only four components are named: Body, Effort, Space, and Shape, which are referred to with the acronym BESS. The relationship among these components is modeled as a tetrahedron, illustrating the multifaceted complexity of adult humans whose movement can be analyzed with any of the components as the primary lens.

Adding a new component to the original four allows us to tease out the implicit notions of time in the system. The fifth component—what we have called *Time*—adds explication of concepts like duration and rhythm. We note that time is foundational in our application to engineering because it allows us to separate duration from "attitude" or "intent," which are topics associated with Effort. These temporal ideas also affect our understanding of phrasing and how we chunk together salience in the temporal dimension (commonly called the "movement segmentation" problem in engineering research). Along with this new component, we suggested a hierarchical relationship between components that is not traditionally part of the presentation of the material of LBMS. We used this hierarchy to suggest new conventions for notating movement in chapter 10.

This new explication also highlights where existing movement theory relies on a shared experience of being human beings on Earth, sharing a built world that is designed for typically upright, bipedal *Homo sapiens*. For example, terms like "gradual" and "abrupt" as used in movement studies actually imply something about both space *and* time. Humans understand this intuitively or automatically because we have gone through many experiences where dramatic spatial changes occur at the same time as rapid

---

Box D.1
Benefits of the Time Component

- It completes the triad of *bodies* moving in *space* and *time*, giving each element of this physical picture their own component of concepts.
- It helps us grapple with experiences of time, like those contained in the embodied exercises in chapter 6 (indeed, these exercises seem to better capture this component than our words can do).
- It aligns with our discussion about shape and effort (particularly shape and effort qualities), relating effort to patterns of a body in time (our *dynamosphere*) and shape to patterns of a body in space (our *kinesphere*).
- It differentiates duration from time effort.
- It enables analysis to distinguish the spatial and temporal aspects of abrupt and gradual change.
- It places "phrasing" inside a dedicated component, aligning with the movement segmentation problem in machine learning and robotics, which may offer insights into movement studies (and vice versa).
- It brings musicality (e.g., rhythm and tempo) into the system more formally.

---

temporal changes—which we describe as an "abrupt" action. But to make a robot move *abruptly*, we need to specify not only that the movement should involve a large change in space, but also that this change must occur over a short duration in time. Thus, it becomes a natural by-product of working with machines to examine human experience in greater detail.

Moreover, humans today use machines that make time more visible than perhaps it was in Rudolf Laban's time. We all carry around what are very sophisticated devices for recording and measuring movement: small computers, mobile phones, or even wearable motion trackers, equipped with cameras, accelerometers, and touch-sensitive screens. These devices allow us to record pictures and videos of ourselves, sharing them instantly with friends. They are designed to estimate how many steps we have taken, how many minutes of exercise we have logged, how many hours of sleep we have gotten, and more. This brings our qualitative experience of time into quantitative focus. Box D.1 offers an incomplete list of how this new Time component aids the broader goals of movement analysis.

Appendix E: Pedagogy and Group Work

In this appendix, we share advice and learnings on teaching this material both inside (section E.1) and outside (section E.2) traditional engineering classrooms settings, as well as embodied exercises best suited for larger groups.

E.1   University Courses at the Intersection of Movement Studies and Technology Design

In the spring semesters of 2017 and 2019, Amy developed a graduate-level research seminar–style course for students in mechanical science and engineering at the University of Illinois at Urbana-Champaign, ME 598: Movement Representation and High-Level Robotic Control. The brief course syllabus included information about the teaching team, office hours, a course description, grading mechanics, reading list, and projected schedule. These five elements, from the 2017 syllabus, are copied here to give educators a potential jumping-off point for developing their own courses.

Course Info

Spring 2017; Seminar: Tuesday 2–5pm in MEL 1208

Prof. Amy LaViers (alaviers@illinois.edu); Office Hours: by appointment in MEL 2113

TA: Umer Huzaifa (mhuzaif2@illinois.edu); Office Hours: by appointment in MEL 1208

Course Description

This interdisciplinary seminar will build on the fact that robots can do basic things. We see them in factories, warehouses, and surveillance where simple, repeatable motions can achieve desired performance. For more complex or dynamic tasks, we need new tools to allow people to create more complex motion programs, ensure their safety, and coordinate among diverse teams. This course will explore a suite of established, more recent, and repurposed tools that are achieving these goals. After this course, students will know how to generate complex behavior in single robots and multiagent systems and will better appreciate the complexity of natural movement from schooling fish to ballet dancers. Emphasis will be placed on where these topics arise in the research projects in the Robotics, Automation, and Dance (RAD) Lab.

The course will have three main components: representation, control, and coordination. Course assignments will include interactive, embodied movement sessions, assigned reading, presentations to fellow students, leading in-class discussions, and developing material for a future course textbook. Topics covered in representation will include embodied movement, movement notation, continuous dynamical systems, discrete transition systems, and graphs. Topics covered in control will include motion tracking, planning, supervisory control, and formal specification. Topics in coordination will include Laplacian dynamics, consensus, leader-follower networks, and formation control. The course will come full circle when we see the relationship among these three topics; the coordination and performance we can achieve on these devices depends on our choices in representation and control strategy. The course prerequisites are courses in at least a few of the following: linear algebra, differential equations, digital logic, Laban Movement Analysis (LMA), Bartenieff Fundamentals (BF), feedback control, and/or networks.

Course Components

- Class participation (30 percent) may be achieved through class attendance with no more than one missed seminar, comments in class that contribute to discussion, helping fellow students understand the course material, finding and reading further resources, and participating in in-class activities, to name a few. Students will write a report at the end of the semester documenting and motivating how they participated in class.

- Assignments (40 percent) will be achieved through completing choreography assignments, assigned reading, presentations to fellow students, leading in-class discussions, and developing material for a future course textbook. Students will turn in a detailed assignment log for this component.

- A final project (30 percent) will be created to express each student's unique takeaway.

Readings

This class will not have a formal textbook, but you will be provided assigned readings from technical papers and books (see below for example list). Additionally, you will go out and find your own material to support your learning in this course.

- *Choreo-graphics: A Comparison of Dance Notation Systems from the Fifteenth Century to the Present.* Hutchinson Guest. Routledge. 1998.
- *Calculus of Variations and Optimal Control Theory: A Concise Introduction.* Liberzon. Princeton University. 2012.
- *Controls and Art: Inquiries That Intersect the Subjective and Objective.* LaViers and Egerstedt (Eds). Springer. 2014.
- *Dance Notations and Robotic Motion.* Laumond and Abe (Eds). Springer. 2015.
- *Dynamic Logic.* Harel and Kozen. MIT Press. 2000.
- *Everybody is a Body.* Studd and Cox. Dog Ear Publishing. 2013.
- *Graph Theoretic Methods in Multiagent Networks.* Mesbahi and Egerstedt. Princeton University Press. 2010.
- *Introduction to Automata Theory, Languages, and Computation.* Hopcroft, Motwani, and Ullman. Pearson. 3rd Ed. 2006.
- *Intro. to Discrete Event Systems.* Cassandras and Lafortune. Springer. 2nd Ed. 2009.
- *Labanotation: The System of Analyzing and Recording Movement.* Hutchinson
- Guest. Routledge. 4th Ed. 2005.
- *Linear Systems Theory.* Hespanha. Princeton University Press. 2009.

- *Making Connections: Total Body Integration Through Bartenieff Fundamentals*. Hackney. Routledge. 2002.
- *Principles of Robot Motion*. Choset, Lynch, Hutchinson, Kantor, Burgard, Kavraki, and Thurn. A Bradford Book: Intelligent Robotics and Autonomous Agents series. 2005.
- *Switching in Systems and Control*. Liberzon. Birkhuser. 2003.
- *Verification and Control of Hybrid Systems: A Symbolic Approach*. Tabuada. Springer.
- 2009.
- *The Vision of Modern Dance: In the Words of Its Creators*. Brown. Princeton Book Company. 1998.

Projected Schedule

Date	Seminar Topic	HW Topic
	– Representation & Expression –	
T Jan. 17	Robotics and movement	What can we learn from dance?
T Jan. 26	Dance (presentations)	"
T Jan. 31	Movement representation	How should we represent movement?
T Feb. 7	Descriptive Writing	"
T Feb. 14	Computable numbers	*no assignment*
	– Coordination & Control –	
T Feb. 21	Graphs / multi-agent systems	Why, how groups?
T Feb. 28	Motif and Labanotation	"
T Mar. 7	Logic and computation / formal methods	How expressive are LTL/Motif/X?
T Mar. 14	Project formulation	"
~~T Mar. 21~~	~~Spring Break, no class~~	
	– Revisiting, Projects, Wrap up –	
T Mar. 28	Project Pitches	Revisit Weeks 1–4
T Apr. 4	Representation and expression	Revisit Weeks 5–8
T Apr. 11	Coordination and control	Project work
T Apr. 18	Project work time (tentative)	"
T Apr. 25	Interconnections and frontiers	"
T May 2	**Final Project Presentations**	*Note: "we" = roboticsists

Figure E.1
Projected Schedule from ME 598 Spring 2017.

Courses similar to Amy's are popping up across institutions of higher learning, such as the following:

- Courses in human-robot-interaction (HRI), design, and human factors are also constantly broadening their subject areas, and teaching the BESST System may well be standard practice one day in the future.

- New courses are constantly be designed and offered in interdisciplinary programs like Arizona State University's School of Arts, Media, and Engineering and similar programs at schools including the University of California, Santa Barbara, Carnegie Mellon University, the University of Colorado Boulder, Rensselaer Polytechnic Institute, Aalborg University, Goldsmiths (University of London), and Princeton University.
- An interdisciplinary course on choreobotics at Brown, taught by Sydney Skybetter in theater arts and performance studies and Stefani Tellex in computer science, was offered in spring 2022.
- Michael Neff, a Certified Laban/Bartenieff Movement Analyst (CLMA) and jointly appointed faculty member in computer science and cinema and digital media at the University of California, Davis, uses LMA to train students in animation.
- In movement studies, faculty such as John Toenjes at the University of Illinois at Urbana-Champaign, Kim Brooks Mata at the University of Virginia, Kate Sicchio at the University of Richmond, and Kate Ladenheim at the University of Maryland use technology within their artistic practices and in courses for dance students.
- Movement analysis certification programs like those being offered at the EMOVE Institute increasingly cite technology design as a key application of the work.

E.2 Movement Workshops on the BESST System

Movement studies is a useful tool for teaching engineers because the body is both the object and subject and both the internal experience and the outward expression are relevant, utilized, explored, and articulated. There is no particular technical virtuosity required to have full access to the material because we are all successful movers. The "ask" is to engage with your body and come to conscious awareness of your choices and experiences in relationship to yourself and to others. Rather than requiring a particular physical performance, such as that associated with dancers, the BESST System helps highlight how any animal moving in the world is already making and perceiving patterns in movement. Everyone brings a unique body, set of movement patterns, and corpus of prior experiences to this work. Unlike in the traditional academy approach to learning, there is no "right" way or "correct" answer for engaging in the movement material. This can be a challenge for the student, who becomes concerned about getting it right because what is "right" is *relative to each unique person's body and body of experience*. The "rightness" comes in allowing oneself to explore, question, sense, attend, and think about what is being experienced, and then also allowing for observing others without judgment.

In this vein, a common way that we open studio-based classes and workshops is by saying, "You are already a competent and capable movement analyst." The somatic strategies and the choreographic technologies that we develop from the BESST System often access improvisational exploration and offer containers of ideas for articulating movement. While there are suggested ways of moving to access the idea and experience, the movement itself is not prescribed; rather, the idea is prescribed. Virtuosity is measured relative to self, not to other, nor to an idealized form. Success is in the mover finding integration, ease, and efficiency of inner intention in relation to outward expression.

Inside such a process, the goal is to raise conscious awareness of movement experience, patterns, habits, and preferences, broadening the bandwidth that we have for sensing, expressing, and describing movement. This increases the ability to observe and interpret movement while recognizing context, prior experience, and bias. As such, the BESST System offers a means to articulate, not only in movement, but also in language—to begin to be able to know what you are doing, what you are perceiving, where it is happening, how it is being expressed, what the purpose or goal of the movement is, and when it is unfolding.

It is important to note that the culture of the classroom and learning environment of this work is quite different than that in a traditional classroom setting. We often work in studio/performance spaces. (No desks, so: sitting on the floor, rolling around on the floor, moving freely without encumbrance in the environment.) This raises the need to be aware of prior experience of all the movers in the room and what the expectations and conventions around being observed, observing, and touching ourselves and each other are. Often we wear special clothes: form-fitting, comfortable clothing, like leggings. This in and of itself can be challenging for the newcomer—potentially creating a sense of vulnerability.

In teaching and practicing this material, we utilize touch as a resource for supporting integration of information with experience. Touch is one way that we can create awareness; just as we can record our movements and then watch the video to see and hear ourselves, using haptic feedback is another way of seeing. Human haptic capacity allows the fine, dexterous manipulation and assembly of objects and our ability to distinguish across a wide variety of surface textures, offering another way to support and sense bodily change (Olsen & McHose, 2004). It is also one of the primary ways in which the differentiation of Self/Other can be experienced. Touch is used to enhance and support our capacity for receiving and responding to sensory information.

In this book, we encourage the reader to use haptic exploration as a resource and a way to tune into not just an internal perspective, but as a way to learn, explore, and form a knowledge of your bodily self. Noticing internal sensations such as the flow of breath or the beat of your heart can be facilitated by putting your hands on your abdomen or over your heart. By using touch to palpate the skeleton, feeling muscles contracting and lengthening and the rise and fall of your breath, an underlying awareness of body structure and function can emerge. In doing so, you learn about yourself. Your body is your laboratory, and touch is a research tool for exploring, learning, supporting and finding new experiences. How we touch ourselves informs our body attitude and is a very useful (but optional) tool for engaging in the embodied exercises offered in this book.

Norms around touch, which are always changing within society at large and differ between cultures and contexts, are different between dance and engineering communities. Thus, in using touch, it is very important to be aware of the need to ask permission before touching another person and also respecting hesitation or resistance to touch. It is also important to allow students to participate even if they do not want to be touched and receive haptic feedback from the instructor, other students, or themselves. This is part of the larger idea that there is no "right" way to engage in embodiment practices.

Engaging with LBMS and learning the BESST taxonomy is a constant process of bringing awareness of the whole to the parts and then returning and integrating the parts to the whole. In other words, it is an ongoing phrase of analysis leading to synthesis by recognizing and exploring the personal and the universal, choice, motivation, and intent, thereby cultivating a conscious awareness of movement and expressivity, both in self and in observing others.

Embodied Exercises

- **Establishing shared values:** One way to navigate a room full of people coming from different backgrounds is to acknowledge these differences and then agree, as a group in a particular moment and place, on shared values. This is often called a "value contract." It can be created by a leader, but it is best when everyone in the room contributes an element to the contract themselves. The leader can demonstrate and steer, making her own additions, to ensure that a wide enough array of values have been established.

  ◦ Begin by designating an object. This could be as simple as a piece of paper or a shared digital document, or something germane to the moment: a watermelon, an old cover to a robot body shell, or other object.

  ◦ Discuss the goals of the group. These could be goals concerning teaching, research, outreach, conflict mediation, or something else.

  ◦ Begin adding values to the object, writing (or typing) each one offered by each member of the group. For example, values for a course like the ones described here could be as follows:

    ▪ Moving together
    ▪ Celebrating our different bodies and prior experiences
    ▪ Laughing with (not at) each other
    ▪ Celebrating everyone's unique movement choices
    ▪ Respecting everyone's right to sit out of an exercise

- **Entering the room:** This exercise demonstrates an easy inroad to movement with a group that may be unfamiliar with the experience of moving together.

  ◦ Begin by standing in a circle. Have each person combine saying his name and doing a movement.

  ◦ Go around the circle accumulating (1, 1,2, 1,2,3, 1,2,3,4, and so on until all names and associated movements have been added).

  ◦ Practice the whole phrase together.

  ◦ Now, have everyone change places in the circle and find the phrase again.

  ◦ Repeat this as many times as the group wants. Are you able to easily reorder the phrase and remember each person's name and movement?

How do you think you are doing this? What we suggest is that you are engaging in movement analysis!

- **Description-based teleoperation:** This exercise asks, "What kinds of words are more useful in translating a movement idea from one body to another?"

  - Sending one person out of the room, set up a simple manipulation task, such as the following:

    - Balancing a tennis ball on a racket
    - Setting a table with multiple pieces of dinnerware (plates, forks, and other utensils)
    - Assembling a simple structure (of a specific shape) out of blocks

  - Bring back the isolated person (who does not know the manipulation task) after tying a blindfold around their eyes so that they cannot see.

  - Elect one or two people who know the task to give verbal cues that allow the blindfolded participant to attempt the manipulation task.

  - What kinds of commands work well? Revisit this exercise after reading all of part II of this book; see if you can identify which commands address the elements of Body, Space, Time, Shape, and Effort from the BESST System.

This version of the task relies on sighted and hearing individuals using sight and sound to accomplish it. Try adapting to other senses, particularly if you have members of your group without full access to vision or audition.

- **Movement specification on distinct bodies:**

  - Use the concepts introduced in this book to create and describe three movement behaviors.

  - Write down your descriptions and *then* enact them in your own body.

  - Pass your descriptions to a colleague to see how they enact them. Watch both your and your partner's sequences to identify commonalities and differences.

  - Compare aspects of the movement through the lens of what is functional (i.e., necessary to complete the action successfully) and what is expressive (i.e., necessary to make the movement seem the same between sequences).

This works well with a large group of people who can contribute to the comparative discussion.

Glossary

A

Abrupt—a type of change (or movement) that utilizes large transformation in its progression in any aspect or aspects of action (e.g., location, rotation, speed, and/or intent).

Accumulation—a type of sequencing that repeats a previous action or set of actions and then adds a new action to the set, which will be included in the next repetition.

Action drive—a mode of movement quality that utilizes eight effort configurations that consist of combinations of weight effort, space effort, and time effort and form the eight basic effort actions (BEAs): float, punch, glide, slash, dab, wring, flick, and press.

Action stroke—the basic unit of symbolic representation of movement used in a vertical motif. It is drawn as a vertical line, the length of which indicates the relative duration of the perceived action it represents, and can be preceded by a pre-sign (or replaced entirely by a symbol that stretches) for more specificity.

Advancing shape quality—complex, three-dimensional change in a form that indicates or expresses a sense of forwardness in the sagittal dimension.

Affinities—relationships between the components of the BESST System that arise from commonalities and trends in human bodily experience, as used to match, or heighten, one another. *See also* "Disaffinity."

Arm circle—the sixth exercise in the Basic Six. It focuses on finding the rotational capacity of the shoulder girdle and arm.

Artificial embodiment—a description of physical capabilities for machines as distinct from human (natural) embodiment, which is analogous and complementary to the term "artificial intelligence."

A scale—a movement scale inside the icosahedron that uses a special type of transverse pathways called "transversals," associated with the feminine gender.

Awake state—a mode of movement quality that utilizes four effort configurations that consist of combinations of space effort and time effort.

Axial movement—a basic body action that does not cause the body to travel through the environment.

Axis of length—the Body Fundamentals (BF) Principle that refers to the sense of verticality in human bodies on Earth.

Axis scales—movement scales inside the icosahedron that use transverse pathways in relation to a specific diagonal.

B

Bartenieff Fundamentals (BF)—the term used to describe much of Irmgard Bartenieff's pioneering work. In conjunction with Laban Movement Analysis (LMA), it forms the contemporary field of Laban/Bartenieff Movement Studies (LBMS) and forms the centroid of the inclusion of somatic practice in the field. *See also* "Body Fundamentals."

Bartenieff Fundamentals Principles (BF Principles)—*See* "Body Fundamentals Principles."

Baseline—a neutral state of the body, formed in context and personal preference, from which movement can differentiate to reveal qualitative and dynamic choices that can express intent.

Basic body actions—a list of movement concepts that offer descriptive power without assigning valued or emotive judgment, divided into axial and locomotor movements, which include change of support, posture, gesture, condense, expand, rotate, vocalize, focus, touch, hold, roll, slide, crawl, walk, run, and jump.

Basic effort actions (BEAs)—the eight configurations of action drive (using the effort factors of weight, space, and time), which are float, punch, glide, slash, dab, wring, flick, and press.

Basic Six—a set of movement ideas, sequences, and patterns developed to get to the essential level of body connectivity and the mover's conscious awareness of patterns of body-level connections. The set consists of thigh lift, sagittal pelvic shift, lateral pelvic shift, body half, knee drop, and arm circle.

BESST System—the codification of concepts in movement studies (especially influenced by the Laban/Bartenieff tradition) presented through the components of Body, Effort, Space, Shape, and Time, which form the acronym BESST.

Body component—the component of the BESST System that addresses "what" is moving.

Body Fundamentals (BF)—the part of the Body component of the BESST System that encompasses Body Fundamental Principles and the Basic Six. Also known, in other texts, as "Bartenieff Fundamentals." *See also* "Bartenieff Fundamentals."

Body Fundamentals Principles (BF Principles)—foundational ideas of body connectivity designed to support clarity in movement intention and experience, which are dynamic alignment, axis of length, body-level phrasing, breath support, core support, rotary support, weight support and shift, developmental pattern support, spatial intent, temporal intent, shape intent, and effort intent. Also known, in other texts, as "Bartenieff Fundamentals Principles."

Body half—the fourth exercise in the Basic Six. It focuses on differentiating the right and left sides of the body by connecting the upper and lower halves on each side.

Body-level phrasing—the Body Fundamentals (BF) Principle that deals with the initiation and sequencing of movement. "Initiation" describes where the movement begins (a spatial location signified by body part), and "sequencing" describes how the movements are ordered in time. *See also* "Sequencing."

Breath support—the Body Fundamentals (BF) Principle highlighting the role of breath (both volitional and unconscious) in human movement. This principle promotes volitional breath to aid in conscious execution and perception of movement.

B scale—a movement scale inside the icosahedron that uses a special type of transverse pathways called "transversals," associated with masculine gender.

C

Canon—a type of sequencing where two movers use the same sequence of actions but shift them in time. One mover waits an amount of time (often a measure when moving to music) before beginning their action. Eventually, both movers move at the same time and follow the same sequence, but it is shifted in time.

Center of gravity—approximately the centroid of the region of the core associated with the lower body, locomotion, and grounding. In humans, it is typically the pelvic girdle, iliac crest, sacrum, and associated abdominal muscles and viscera, which almost always includes the true "center of mass" and "center of gravity" of the body (in the sense that the terms are used in physics).

Center of levity—approximately the centroid of the region of the core associated with the upper body, communication, and lifting. In humans, it is typically the skull, collarbone, sternum, and associated abdominal muscles and viscera.

Change of support—a change in shape and/or center of mass of a mover and, thus, a resulting change in contact between body and environment. It can be considered as the most generic basic body action, as any action in the body will inherently cause a change of shape and/or center of mass.

Choreographic principles—ideas and knowledge used to guide the process and analysis of choreography.

Choreographic technologies—the use of choreography in a practical pursuit, such as developing machine behavior.

Choreography—an arrangement of movement in time and space or the process of designing movement for a particular artistic or practical goal.

Clock column—the leftmost column of the CPMMPT staff, which, optionally, indicates absolute duration, meter, or tempo of notated movements and phrases. *See also* "CPMMPT staff."

Component constellation—a movement notation style that visually organizes the elements of the BESST System present in a particular movement observation that employs the hierarchical model of BESST. It uses a triangle staff and reveals elements essential to the observation and their relationship in movement theory, but it does not reveal order, frequency, or duration. *See also* "Triangle staff."

Condensing—one of the contrasting polarities (the opposite of indulging) used in the BESST System that refers to ideas of making, fighting, and going against. For example, each effort factor has a condensing polarity. *See also* "Indulging."

Constellation motif—a movement notation style that indicates a group of "ingredients" that are present in the movement event. It uses a constellation staff and reveals elements essential to the event but not order, frequency, or duration. *See also* "Constellation staff."

Constellation staff—the geometric pattern used to arrange symbols in constellation motif, comprised of four dots around unordered symbols. *See also* "Constellation motif."

Core—the "center" of a moving body. This is a flexible, contextual term used to describe an aggregation of body parts that are being viewed as distinct from the body parts that create a limb. For example, in the human body, this term may refer to the spine, including the head and tail, and associated muscles, bones, and viscera, but sometimes the head and/or tail are excluded from this region, as they can also be viewed as fifth and sixth limbs that have the possibility for articulation. *See also* "Limb."

Core support—the Body Fundamentals (BF) Principle highlighting the role of the core in human movement. The trunk (or core) of the body, which includes the spine, pelvis, shoulder girdle, and internal organs, is the place where movement can originate, activate, and support our successful negotiation of our relationship to gravity.

Coronasphere—the region of space around breathing bodies in which their movement, especially of their breath, can occur or be perceived.

CPMMPT staff—more detailed version of the vertical staff (or motif) for movement notation, delineated by Clock, Part, Main Action, Modification, Phrasing, and Theme columns, developed for applications of movement analysis to machines. *See also* "Clock, Part, Main Action, Modification, Phrasing, and Theme column."

D

Dab-wring diagonal—a spatial form in the cube, between two vertices that are opposite one another through the longest span of the shape; with right side leading, it is traced between left-back-high, which is affined with the basic effort action (BEA) of dab, and right-forward-low, which is affined with the BEA of wring.

Defense scale—a movement scale inside the octahedron that uses all peripheral pathways.

Developmental pattern support—the Body Fundamentals (BF) Principle that deals with the predicated, patterned progression of limb/core integration in human motor development. This principle involves both accessing all these patterns, as adult movers, and returning to and isolating earlier patterns in human development to increase a mover's ability to support complex movement in the environment.

Diagonal scale—a movement scale inside the cube that uses alternating central and peripheral pathways. *See also* "Dab-wring, Flick-press, Float-punch, and Glide-slash diagonal."

Dimensional scale—a movement scale inside the octahedron that uses central and, in one variant, alternating central and peripheral pathways.

Directional shape change—a mode of shape change that describes actions that connect or bridge to the environment using either an arc-like or spoke-like style.

Disaffinity—relationships between the components of the BESST System that arise from commonalities and trends in human bodily experience, as used in contrast to one another. *See also* "Affinities."

Door plane (vertical plane)—the plane bisecting the human form that is perpendicular to the typical direction of forward travel (or sagittal dimension); it is called the "coronal plane" in anatomy. Associating the plane with a door brings forth an image of stability and flatness of a hinged door.

Dream state—a mode of movement quality that utilizes four effort configurations that consist of combinations of weight effort and flow effort.

Drives—effort configurations with three factors present, which describe heightened, rare, and especially rich moments of movement quality.

Duration—how long an action lasts (an interchangeable concept with the speed of an action). Duration may be relative to other actions or an absolute measurement; it is explicated as a formal element in the BESST System to better describe affinities between the length of an action, phrasing, and time effort.

Dynamic alignment—the Body Fundamentals (BF) Principle that develops the idea of skeletal connections and interconnections to support access to kinetic chains of movement, framing a key idea that every and any motion in the body affects all of the body.

Dynamosphere—the temporal "space" of a moving body, which captures the qualitative, subjective aspect of the Time component; analogous to the relationship between the kinesphere and the Space component.

E

Echoing—a type of sequencing strategy where one mover performs an action and then a second mover performs the same action, delayed in time, as an "echo" of the first mover.

Effort component—the component of the BESST System that addresses "how" movement is happening.

Effort factors—constructs used to model and notate motion quality, with respect to the body in time. The relationship between the four factors, weight, flow, space, and time is visually represented on the Effort graph.

Effort intent—the Body Fundamentals (BF) Principle that deals with a mover's motivation as manifest in movement quality, explicitly related to the Effort component, but foregrounding the bodily, physical experience that creates these more complex ideas.

Enclosing shape quality—a complex, three-dimensional change in form that indicates or expresses a sense of closing in the horizontal dimension (e.g., the right side of the body closing toward the left).

Exertion/Recuperation theme—an overarching theme of the BESST System that addresses the inextricable interrelationship of work and rest, especially in the perception of moving bodies.

Expressive robotics—an approach to robotics that investigates and develops machine movement foregrounding expression over function. Traditionally, the field has foregrounded function over expression, due to the large application of robotics in manufacturing. Often, knowledge from dance (e.g., choreography), as well as other physical practices and the arts (e.g., tai chi and sculpture), are used in this approach. In this book, "robot" is viewed broadly as a term for especially advanced or adaptive technology that causes movement in the environment (of human and/or machine bodies), and "robotics" is viewed as a field in need of reintegrating function and expression to better serve humans with machines in all applications—by employing expertise from movement studies.

F

Flick-press diagonal—a spatial form in the cube, between two vertices that are opposite one another through the longest span of the shape; with right side leading, it is traced between right-back-high, which is affined with the basic effort action (BEA) of flick, and left-forward-low, which is affined with the BEA of press.

Float-punch diagonal—a spatial form in the cube between two vertices that are opposite one another through the longest span of the shape; with right side leading, it is traced between right-forward-high, which is affined with the basic effort action (BEA) of float, and left-back-low, which is affined with the BEA of punch.

Flow—a description of movement experience that refers to both release and control and foregrounds the connection to the environment and others in it.

Flow effort—the effort factor that arises out of the fundamental body experience of flow-sensing and describes the expression of progression and feeling. The condensing polarity is bound, and the indulging polarity is free.

Flow-sensing—actively noting and engaging with the bodily experience of flow.

Function/Expression theme—an overarching theme of the BESST System that addresses the inextricable interrelationship of functionality and expressivity, especially in the perception of moving bodies.

G

Girdle scale—a movement scale inside the icosahedron that uses peripheral pathways, orbiting a specific diagonal.

Glide-slash diagonal—a spatial form in the cube between two vertices that are opposite one another through the longest span of the shape; with right side leading, it is traced between left-forward-high, which is affined with the basic effort action (BEA) of glide, and right-back-low, which is affined with the BEA of slash.

Gradual—a type of change (or movement) that utilizes small transformation in its progression in any aspect or aspects of action (e.g., location, rotation, speed, and/or intent).

H

Horizontal motif—a movement notation style that indicates essential elements and includes information about the order in which the elements occur. Symbols are presented left to right in order of their appearance. It uses a horizontal staff and reveals basic elements of the event and their ordering, but it does not specify duration. *See also* "Horizontal staff."

Horizontal staff—the geometric pattern used to arrange symbols in horizontal motif, comprised of two bars that precede and follow a series of ordered symbols that are read left to right. *See also* "Horizontal motif."

I

Indulging—one of the contrasting polarities (the opposite of condensing) used in the BESST System that refers to ideas of unmaking, allowing, and going with. For example, each effort factor has a indulging polarity. *See also* "Condensing."

Inner/Outer theme—an overarching theme of the BESST System that addresses the inextricable interrelationship of internal and external experience, especially in the perception of moving bodies.

Inner shaping—the deformation of the body's innersphere that supports shaping, leveraged most centrally during breathing and differentiated in each of the three ordinal dimensions through the terms: lengthening/shortening (vertical dimension), bulging/hollowing (sagittal dimension), and widening/narrowing (horizontal dimension).

Inner space—*See* "Innersphere."

Innersphere—the region of space within articulated bodies in which their movement, especially of their internal organs and viscera, can occur or be perceived. Also referred to as "inner space."

K

Kinesphere—the region of space around articulated bodies in which their movement, especially of their limbs, including the head and tail, can occur or be perceived; analogous to the relationship between the dynamosphere and the Time component.

Kinesthetic attunement—the process of taking the movement of a foreign body into one's own personal body to support more nuanced observation.

Kinetic chain—a possible connection between bony landmarks of the body; typically used in teaching and rehearsing movement phrases. For example, focusing on the series of bodily elements and actions between the head and the tail (or coccyx) can help engage the core and spine in achieving balance and bodily control.

Knee drop—the fifth exercise in the Basic Six. It focuses on finding a twist of the lower unit in relationship to the upper unit through rotation.

L

Laban/Bartenieff Movement Studies (LBMS)—a contemporary term meant to describe the field that has emerged from Laban Movement Analysis (LMA) and Bartenieff Fundamentals (BF). It sits at the nexus of interior, body-based experience (somatics) and exterior, performative pursuits (choreography).

Laban Movement Analysis (LMA)—a term sometimes used interchangeably with "Laban/Bartenieff Movement Studies (LBMS)," but that emphasizes Rudolf Laban's contributions to the field, which are more rooted in choreography than somatics.

Labanotation—a movement notation style developed by Rudolf Laban that uses a staff based on human anatomy and moves from the bottom to the top of the page.

Lateral pelvic shift—the third exercise in the Basic Six. It focuses on a sideways mobilization of the pelvis in a core-to-limb pattern.

Level—a way to divide or organize the kinesphere, marked as relative distance from the ground.

Limb—the "extendable edges" of a moving body. This is a flexible, contextual term used to describe body parts that capable of direct interaction with the environment (e.g., for manipulation or gesture). For example, in the human body, this term may refer to the arms or legs and associated muscles, bones, and viscera, but sometimes the head and tail are viewed as fifth and sixth limbs that have the possibility for articulation. *See also* "Core."

Locomotor movement—a basic body action aimed at moving a body to a new location in the environment.

M

Main action column—the left central column of the CPMMPT staff. Symbols within it indicate the main action of a movement sequence, using action stroke, basic body actions, and spatial pulls, revealing relative duration (e.g., as embedded in the length of an action stroke). *See also* "CPMMPT staff."

Meter—a type of temporal (especially rhythmic) structure (e.g., 4/4 meter uses an even count of 4 or 8 to delineate time).

Mobile state—a mode of movement quality that utilizes four effort configurations that consist of combinations of flow effort and time effort.

Modes of shape change—the three specific and distinct ways the body's form changes in relationship to self and the environment: shape flow, directional shape change, and shaping.

Modification column—the right central column of the CPMMPT staff. It provides space for symbols that modify the symbols in the Main Action column, giving more detail to the nature of the notated movement. *See also* "CPMMPT staff."

Motif—the type of movement notation used in the BESST System. Unlike systems like Labanotation, which uses many of the same symbols but aims for strict recording rooted in human anatomy, motif is used to describe the essence of a movement phrase or idea.

Motion factor—*See* "Effort factors."

Movement—a perceived change (as distinct from the motion that all matter and energy are constantly undergoing). This change may be perceived by any of the human senses and is typically the result of several working in concert.

Movement analysis—the process of acquiring data and making observations of a moving body to identify patterns. The practice is common to the Laban/Bartenieff tradition, as well as the broader field of dance, and it is also used in biology, kinesiology, and biomechanics.

Movement platforms—natural or artificial bodies of any morphology onto which movement is specified and/or choreographed.

Movement scale—a sequence of spatial pulls inside the kinesphere modeled as various geometric ideals enacted by a mover to practice a particular approach to kinesphere, explore new patterns of movement, and find spatial harmony in their choices.

Movement studies—a broad field that encompasses both theory and practice of bodily movement from an interior and exterior lens, consisting of three main components, somatics, choreography, and notation, and drawing from the narrower academic fields of dance, kinesiology, performance studies, biomechanics, and robotics.

N

Nonsimultaneous sequencing—movement events that happen before or after each other, not at the same time (in series, not in parallel).

Notation—the use of symbols to denote and represent abstract ideas (e.g., music, mathematics, movement), that are often used in translating ideas across bodies or contexts (e.g., musical instruments, models, and human dancers, respectively).

Notational abstraction—a broad idea about movement that can be used to find similarity across movement instances, and thus drive symbology to notate movement.

O

Out-of-step—a type of sequencing strategy where each mover performs the same sequence, but not exactly at the same time as (i.e., not in unison with) every other mover.

P

Passion drive—a mode of movement quality that utilizes eight effort configurations that consist of combinations of weight effort, time effort, and flow effort, associated with having no sense of attention or thinking. Passion drive has eight effort configurations, such as passion drive float, where indirect space effort in the body effort action (BEA) of float is replaced with free flow effort.

Pathway—a way to divide or organize movement through the kinesphere marked by the manner the movement progresses between the center and distal reaches of the kinesphere, which can be central (passing through the center), peripheral (staying at the distal edge), or transverse (going in between the edge and center, avoiding both).

Patterns of body organization—body relationships that emerge from human six-limbedness and preference for two of these limbs (head and tail) to be more significant. These relationships are: radial symmetry, spinal, core/distal, head/tail, upper/lower halves, right/left halves, and cross-lateral.

Phrase column—the second column from the right of the CPMMPT staff, which optionally indicates the types of phrasing in the sequence using phrasing bows. *See also* "CPMMPT staff."

Phrasing—the grouping of movements that belong together in a single temporal stream (analogous to the phrasing of notes in music).

Platform column—the second column from the left of the CPMMPT staff, which, optionally, indicates a body part for a specific platform, allowing notators to intentionally create assignments to accommodate different platform morphologies with greater specificity. *See also* "CPMMPT staff."

Platform-invariant representation—a movement instruction or command that is not specific to a particular movement platform (e.g., modern music notation).

Platform-specific representation—a movement instruction or command that is specific to a particular movement platform (e.g., tablature).

Pre-sign—a symbol from the BESST System used to modify an action stroke in a vertical motif. Pre-signs are drawn before (below) the action stroke, and a small linking bow is drawn to connect them to their associated action stroke. This allows the action stroke length to reveal the relative duration of symbols that do not stretch.

Primary patterns of shape change—describe the body's changing form in relationship to the environment, which can be convex/concave or gathering/scattering. Convex/concave focuses on the core (rather than the limbs) and describes the body opening and closing relative to the environment, respectively. Gathering/scattering focuses on the limbs (rather than the core) and describes the body taking and giving relative to the environment, respectively.

Primary scale—a movement scale inside the icosahedron that uses peripheral pathways around a specific diagonal.

R

Reach space—a way to divide or organize the kinesphere, either near, mid, or far, marked as the relative distance from the mover's core, as accessed by the limbs.

Remote state—a mode of movement quality that utilizes four effort configurations that consist of combinations of flow effort and space effort; it is called "far state" in some texts.

Repetition—a type of sequencing strategy where an action or a set of actions is repeated in time, often creating emphasis or desensitization.

Retreating shape quality—a complex, three-dimensional change in form that indicates or expresses a sense of moving backward in the sagittal dimension.

Retrograde—a sequencing strategy where actions are performed backward in time (trying to imitate the effect of rewinding a video).

Reversal—a sequencing strategy that transforms a sequence of actions in reverse order.

Rhythm—a structure for perceptually marking time.

Rhythm state—a mode of movement quality that utilizes four effort configurations that consist of combinations of weight effort and time effort; it is called "near state" in some texts.

Rising shape quality—a complex, three-dimensional change in form that indicates or expresses a sense of moving upward in the vertical dimension.

Rotary support—the Body Fundamentals (BF) Principle that deals with the role of body part rotation in human movement. Access to complex, three-dimensional movement is supported by the rotary capacity of human joints, which movers can cultivate and articulate.

S

Sagittal pelvic shift—the second exercise in the Basic Six. It focuses on shifting the pelvis forward and slightly upward in a core-to-limb pattern.

Self/Other theme—an overarching theme of the BESST System that addresses the inextricable interrelationship of what is part of oneself and what is not, especially in the perception of moving bodies.

Sequencing—how movement events are ordered. This can be done in a manner that is simultaneous or two types of nonsimultaneous: sequential or successive. *See also* "Body-level phrasing."

Sequential sequencing—a type of nonsimultaneous sequencing where actions may occur in nonadjacent body parts.

Shape component—the component of the BESST System that addresses "for whom" the movement is happening.

Shape flow—a mode of shape change that describes actions that involve change in form of a mover in response to himself or herself (rather than in response to the larger environment outside). Breath is used extensively in this process, and changes associated with the inhale are often described as "growing," while changes associated with the exhale are described as "shrinking."

Shape intent—the Body Fundamentals (BF) Principle that deals with a mover's change in form, both his or her own form and the form in relationship to the environment. It is explicitly related to the Shape component but foregrounds the bodily, physical experience.

Shape quality—a construct used to model and notate motion quality with respect to the body in space, which are visually represented in the Shape quality graph.

Shaping—a mode of shape change that describes actions that accommodate or adapt to the environment in a complex manner, as in manipulation.

Simultaneous sequencing—a movement ordering in which one or more actions occur at the same time.

Sinking shape quality—a complex, three-dimensional change in form that indicates or expresses a sense of moving downward in the vertical dimension.

Somatic practice—the regular and applied use of the field of somatics to develop bodily fluency.

Somatics—a field that uses the body as a site for investigation and knowledge development, particularly with an interior, sensorial focus.

Somatic strategies—the use of somatics in a practical pursuit, such as developing machine behavior.

Space component—the component of the BESST System that addresses "where" movement is happening.

Space effort—the effort factor that arises through our senses and the bodily ways that we focus on the environment, and describes the expression of attention and thought. The condensing polarity is direct, and the indulging polarity is indirect (also called "flexible" or "broad," respectively, in some texts).

Spatial intent—the Body Fundamentals (BF) Principle that deals with a mover's spatial goals in his or her movements, explicitly related to the Space component but foregrounding the bodily, physical experience that creates these more complex ideas.

Spatial pulls—abstractions that engage a mover with the kinesphere, typically associated with the vertices of Platonic solids (e.g., the cube and icosahedron). Rather than destinations where the mover may arrive, these are ongoing or asymptotic ideas about energy and intention that are used to continuously expand the mover's conception of what the body can do.

Speed—*See* "Duration."

Spell drive—a mode of movement quality that utilizes eight effort configurations that consist of combinations of weight effort, space effort, and flow effort, associated with having no sense of commitment or decision. Spell drive has eight effort configurations, such as spell drive float, where sustained time effort in the basic effort action (BEA) of float is replaced with free flow effort.

Spreading shape quality—a complex, three-dimensional change in form that indicates or expresses a sense of opening in the horizontal dimension (e.g., the right side of the body opening to the right).

Stability/Mobility theme—an overarching theme of the BESST System that addresses the inextricable interrelationship of stability and mobility, especially in the perception of moving bodies.

Stable state—a mode of movement quality that utilizes four effort configurations that consist of combinations of weight effort and space effort.

States—effort configurations with two factors present, which are commonly occurring moments of movement quality and texture.

Steeple—a way of phrasing a movement scale: a series of two pathways that are deflections of the same diagonal that is experienced as a pointed, two-part experience of a trio of spatial pulls. *See also* "Volute."

Still shape forms—static patterns in shape, which include pin, ball, wall, tetrahedron, and screw.

Successive sequencing—a type of nonsimultaneous sequencing in which actions sequence through adjacent body parts.

T

Table plane (horizontal plane)—the plane bisecting the human form that is parallel to the ground; it is called the "transverse plane" in anatomy. Associating the plane with a table brings forward the image of bodies reaching from side to side across a table.

Tempo—the rate of temporal delineation, such as beats per minute.

Temporal intent—the Body Fundamentals (BF) Principle that deals with a mover's temporal goals in his or her movements, explicitly related to the Time component but foregrounding the bodily, physical experience that creates these more complex ideas.

Time component—the component of the BESST System that addresses "when" movement happens.

Time effort—the effort factor that arises from our relationship to chronological time and our attitude toward it and describes the expression of commitment and decision. The condensing polarity is sudden (also called "quick" in some texts), and the indulging polarity is sustained.

Time quality—describes the subjective, qualitative aspect of the passage of time between two polarities, gradual (lingering, prolonged, ongoing, and endless) and abrupt (instantaneous, immediate, rapid, and stopped).

Theme and variation—a sequencing strategy that involves riffing on a movement theme, often established through repetition, creating a slight variation in the form of a new movement.

Theme column—the rightmost column of the CPMMPT staff, which, optionally, indicates the theme, or larger idea, of the movement sequence. *See also* "CPMMPT staff."

Thigh lift—the first exercise in the Basic Six. It focuses on hip flexion and the ilio-femoral relationship of lower limb to core.

Transformation drives—the three effort drives that are combinations of flow effort with two other factors (either weight effort, space effort, or time effort), which are: passion, spell, and vision drive.

Transversal—a transverse pathway between spatial pulls of two tensions that move in a special way between the sagittal, vertical, and horizontal planes: it must transit from one plane to another while passing through the third one. These pathways require notably complex bodily accommodation.

Triangle staff—the geometric pattern used to arrange symbols in a component constellation, a style of motif, comprised of an inverted triangle with divisions for each component of the BESST System as well as notes about the observer and context. *See also* "Component constellation."

U

Unison—a type of sequencing across multiple bodies in which the bodies are (perceived as) doing the same thing at the same time.

V

Vertical motif—a movement notation style that uses a vertical staff (written from the bottom of the page to the top) and reveals essential elements, order of occurrence, and relative duration of movement events. It also allows the modification of primary actions for more detailed descriptions of events than other motifs. Multiple modifiers in addition to phrasing and theme bows can be used to reveal emphasis, tone, thematic information, and more. *See also* "Vertical staff."

Vertical staff—the geometric pattern used to arrange symbols in vertical motif, comprised of two bars that precede and follow a series of ordered symbols that are read bottom to top, extending to reveal relative duration. *See also* "Vertical motif."

Vision drive—a mode of movement quality that utilizes eight effort configurations that consist of combinations of space effort, time effort, and flow effort, associated with having no sense of intention or sensing. It has eight effort configurations, such as vision drive float, where light weight effort in the basic effort action (BEA) of float is replaced with free flow effort.

Volute—a way of phrasing a movement scale: a sequence of two pathways that relate to different diagonals that produces a motion between three spatial pulls that is sweeping and rounded. *See also* "Steeple."

W

Weight—a description of movement experience that refers to both active and passive engagements of mass in gravity and foregrounds the connection to and within the body.

Weight effort—the effort factor that arises from the experience of weight-sensing and describes the expression of intention and sensation. The condensing polarity is strong, and the indulging polarity is light. (Some texts also describe a passive weight expression through the terms "heavy" and "limp," respectively.)

Weight-sensing—actively noting and engaging with the bodily experience of weight; often created through small, vibratory perturbations (or jiggles) of the body.

Weight support and shift—the Body Fundamentals (BF) Principle that deals with the role of mass acting under gravity in human movement and is a key concept in locomotion.

Wheel plane (sagittal plane)—the plane bisecting the human form that is parallel to the direction of travel used when locomoting forward; the same term, "sagittal," is used in anatomy. It is associated with a wheel to bring forth the image of mobility and travel.

Z

Zone—a way to divide or organize the kinesphere, marked by the sagittal, vertical, and horizontal planes, which create regions that are above and below, to the front and back, and on both sides of the body.

Notes

Introduction

1. This was NSF Award 1701295, a collaborative grant between the University of Illinois at Urbana-Champaign (where the RAD Lab moved in 2015) and Tulsa University, where a team worked in parallel to our modeling, simulation, and movement design efforts to develop a mechanism in hardware.

2. Amy is a coauthor of five of these.

3. Details of this course and excerpts of its syllabus are included in appendix E.

4. Likewise, there will be little mention of "body language"—a concept of human movement that does not fully capture the ubiquity of movement (notice that vocalization is itself a bodily movement) and the vital roles of context and personal experience in meaning-making.

5. This book does not offer a comprehensive review of any particular field or researcher, including either of the authors. It is primarily a reference on movement studies that is especially amenable to use for research, development, and teaching inside science, technology, engineering, and mathematics (STEM) fields.

6. Such an example was created as a stellar final project in Amy's graduate-level mechanical engineering research seminar in spring 2019 by Reika McNish, a graduate student in kinesiology with an undergraduate degree in dance.

Chapter 1

1. Section 4.6 of chapter 4 presents a formal discussion of these concepts.

2. These definitions come from Oxford Languages (https://www.oed.com/; emphasis added).

3. We will discuss these seemingly opposite ideas as "dualities" (defined in section 3.2 in chapter 3) that have an interlocking, nonlinear relationship—and a

particularly essential role in the system of movement description and analysis presented in part II.

Chapter 2

1. In chapter 10, however, we suggest that it is not so simple to identify whether two movement phenomena are "the same"—in fact, it requires a system of abstraction like the one we introduce in this book.

2. See more about this idea in the Inner/Outer theme introduced in section 3.2 in chapter 3.

3. Associated symbols for certain movement ideas will be introduced throughout part II of the book and reexamined inside a notational system in chapter 10.

4. Section 5.4 in chapter 5 presents an early attempt to do so through movement notation (Jang Sher et al., 2019). Computer programs are also written for particular machines, but there is much more progress toward interoperability in this field: you can often install old software on new computer models, provided that they use the same operating systems and satisfy certain minimum hardware constraints (LaViers, 2019b).

Chapter 3

1. Sometimes the word "system" is used in place of "studies" as in (Studd & Cox, 2013/2020).

2. The Laban/Bartenieff Institute of Movement Studies (LIMS) was founded in 1978 (first as the Laban Institute of Movement Studies) and began codifying a system for movement analysis while offering a rigorous training (520 hours) that led to the degree of Certified Movement Analyst (CMA). Both authors of this book earned their CMA degrees at training programs with LIMS, which is based in New York, New York. Currently, other certification programs include Integrated Movement Studies (IMS) on the West Coast of the US, which trains Certified Laban/Bartenieff Movement Analysts (CLMAs); EMOVE in Europe and WholeMovement in the US and Europe, which certify Laban/Bartenieff Movement Analysts (LBMAs); and Trinity Laban in the UK, which offers a traditional dance degree rooted in this approach.

3. Instead of using LMA, BF, or LBMS, which include Laban and Bartenieff's names, to describe the contemporary system that derived from their work, we will use the term "the BESST System." Many texts will capitalize every term in the system; we will only capitalize the five components and five themes named here when they refer to that element of the system itself (i.e., when they are proper nouns), as opposed to referring to a specific movement observation (i.e., when they are a general property, like energy or entropy).

4. The term "dualism" is also used in philosophy to describe the idea, associated with René Descartes, that the mind and body are separate entities. In fact, this is quite the opposite of the idea of duality in LBMS, which posits that the mind is the body and, simultaneously, the body is the mind. In our view, this is consistent with the nondualistic approach used in somaesthetic design practice (Höök et al., 2021). It is an unfortunate example of using the same word to mean opposite things in different disciplines. Despite the use of the term in design practices and philosophy, we choose to stick with the usage that is consistent with movement analyst training programs. In our view, this usage is consistent with the idea as employed in mathematics, where Möbius transformations, for example, convert a continuous time description to discrete time, creating two different representations of the same idea.

Chapter 4

1. See case study 4 in box 9.4 in chapter 9 for an example of preferred direction for action and sensing (face) applied to machine design.

2. See case study 2 in box 9.2 in chapter 9 for an example of how a consideration of proximal joints across human and robot bodies can be applied to machine design.

3. The terms "proximal" and "distal" can also be used to indicate the relative positions of anatomical parts; for example, the elbow joint is distal to the shoulder joint.

4. Here, we partially adopt the symbols introduced by Studd and Cox (2013/2020) but swap out the backward Effort symbol for a Body symbol to reflect the use of these symbols within the Body component.

5. Breath symbols developed by Curtis Stedge (2017).

6. See case study 5 in box 9.5 in chapter 9 for an example of breath being applied to machine design.

7. See case study 3 in box 9.3 in chapter 9 for an example of weight- and flow-sensing applied to machine design.

8. See case study 2 in box 9.2 in chapter 9 for an example of weight-sensing applied to machine design.

9. See case study 1 in box 9.1 in chapter 9 for an example of the basic body actions of posture, gesture, focus, rotation, and touch being applied to machine design.

10. See case study 5 in box 9.5 in chapter 9 for an example of the basic body actions of posture and gesture being applied to machine design.

11. See case study 2 in box 9.2 in chapter 9 for an example of the basic body action of rotation being applied to machine design.

12. See case study 4 in box 9.4 and case study 5 in box 9.5 in chapter 9 for examples of the basic body action of vocalize being applied to machine design.

13. See case study 3 in box 9.3 in chapter 9 for an example of the basic body action of touch being applied to machine design.

14. See case study 4 in box 9.4 in chapter 9 for an example of the basic body action of travel being applied to machine design.

15. See case study 2 in box 9.2 in chapter 9 for an example of axis of length being applied to machine design.

16. Chapter 6 discusses an affinity between the Body and Time components in connection with this term.

17. See case study 5 in box 9.5 in chapter 9 for an example of breath support being applied to machine design.

18. See case study 2 in box 9.2 in chapter 9 for an example of weight support and shift being applied to machine design.

Chapter 5

1. Bringing up images of the sun's coronasphere, this idea was first suggested in the context of the coronavirus pandemic that began in 2019 by Elswit (2020 and 2021).

2. See case study 2 in box 9.2 in chapter 9 for an example of kinesphere being applied to machine design.

3. See case study 4 in box 9.4 in chapter 9 for an example of zones being applied to machine design.

4. See case study 1 in box 9.1 in chapter 9 for an example of mid-reach space being applied to machine design.

5. See case study 2 in box 9.2 in chapter 9 for an example of high and low being applied to machine design.

6. Often called "Place Middle"; likewise, "high" and "low" are often called "Place High" and "Place Low," especially in Labanotation and older texts. In dropping this proper noun style of reference, we are honoring that these spatial pulls are not specific places. They vary across bodies—and on a given body, across time and space. Place Middle, usually a neutral standing posture, for one person is different for another, and even for that first person after a night of sleeping in a cramped position.

7. See case study 1 in box 9.1 in chapter 9 for an example of spatial pulls being applied to machine design.

8. See case study 4 in box 9.4 in chapter 9 for an example of this theme being applied to machine design.

Chapter 6

1. We include ideas here that have often been implicit in Body, Effort, Space, and Shape, like duration, tempo, and perceived passage of time, but not given their own treatment. See appendix D for a longer discussion of this change.

2. We align our taxonomy with that of Niebles & Fei-Fei (2007) and Fanti (2008).

3. See case study 2 in box 9.2 in chapter 9 for an example of tempo and rhythm being applied to machine design.

4. See case study 3 in box 9.3 in chapter 9 for an example of meter being applied to machine design.

5. An additional element, "preparation," is sometimes listed before initiation (Wahl, 2019).

6. See case study 5 in box 9.5 in chapter 9 for an example of impactive phrasing being applied to machine design.

7. See case study 3 in box 9.3 in chapter 9 for an example of vibratory phrasing being applied to machine design.

8. See case study 3 in box 9.3 in chapter 9 for an example of this theme being applied to machine design.

9. Laban's notion of "harmony of movement" is, most broadly, about balance (Laban, 1966, p. 195).

Chapter 7

1. See case study 4 in box 9.4 in chapter 9 for an example of still shape forms being applied to machine design.

2. Many texts classify this as a basic body action, but we follow the second edition of Studd and Cox (2013/2020), placing this idea inside Shape.

3. See case study 3 in box 9.3 in chapter 9 for an example of shape flow being applied to machine design.

4. See case study 2 in box 9.2 in chapter 9 for an example of directional shape change being applied to machine design.

5. See case study 1 in box 9.1 in chapter 9 for an example of a spoke-like directional mode of shape change being applied to machine design.

6. Some texts refer to this as "carving."

7. See case study 2 in box 9.2 in chapter 9 for an example of rising and sinking being applied to machine design.

8. To date, the video had 25.8 million views across two different postings on YouTube (one by the Boston Dynamics channel at https://www.youtube.com/watch?v=cNZPRsrwumQ and one by user olinerd at https://www.youtube.com/watch?v=W1czBcnX1Ww).

Chapter 8

1. Merriam-Webster dictionary accessed at https://merriam-webster.com/dictionary/.

2. Effort factors are also called "motion factors" (Laban & Lawrence, 1959).

3. See case study 5 in box 9.5 in chapter 9 for an example of flow effort being applied to machine design.

4. See case study 1 in box 9.1 in chapter 9 for an example of direct space effort being applied to machine design.

5. Indirect space effort is also called "flexible" (Newlove & Dalby, 2004) and "broad" (Studd & Cox, 2013/2020) in some texts.

6. Sudden time effort is also called "quick" in some texts (Newlove & Dalby, 2004).

7. See case study 2 in box 9.2 in chapter 9 for an example of the states associated with passion drive being applied to machine design.

8. Remote state is also referred to as "far state" (Laban & Lawrence, 1959).

9. Rhythm state is also referred to as "near state" (Laban & Lawrence, 1959).

10. See case study 3 in box 9.3 in chapter 9 for an example of rhythm state being applied to machine design.

11. See case study 5 in box 9.5 in chapter 9 for an example of awake state being applied to machine design.

12. See figures 10.6, 10.7, and 10.8 in chapter 10 and the surrounding description of sequences A and B for an example of the Effort-Space affinity in action.

13. See case study 2 in box 9.2 in chapter 9 for an example of the affinity between Space and Shape being applied to machine design.

Chapter 9

1. See the process in box 10.6 of chapter 10 for more insight into how this case study, as well as the ones that follow, were completed.

2. We adopt the usage of Bartneck et al. (2020, p. 87), who used "imitation" to mean conscious mirroring and "mimicry" to refer to unconscious mirroring, both of which are thought to be important in HRI and nature.

Chapter 10

1. Some use a different convention, combining the length of the pre-sign and action stroke to reveal relative duration.

2. A body part attached to an action stroke as a pre-sign indicates an action *using that body part* for the relative duration revealed by the length of the action stroke.

3. Some contend that qualities with sudden time effort have an inherent, associated dynamic, and therefore duration does not need to be specified. In this convention, the notator would not use an action stroke in conjunction with the symbols for these effort configurations.

4. In your own body, this step may be rather immediate: just stand up and move. However, when prototyping a machine, this step—and the process of changing how the device works—may be quite time intensive. Work with as many low-fidelity (lo-fi) prototyping materials (e.g., pipe cleaners, cardboard, clay, sketches, or even simulations) to encourage lots of iteration.

Appendix B

1. Exploring scales in other forms, such as the tetrahedron and dodecahedron, is an active area of research in the movement studies community.

References

Abe, N., Laumond, J. P., Salaris, P., & Levillain, F. (2017). On the use of dance notation systems to generate movements in humanoid robots: The utility of Laban notation in robotics. *Social Science Information, 56*(2), 328–344.

Afolabi, O., Driggs-Campbell, K., Dong, R., Kochenderfer, M. J., & Sastry, S. S. (2018, October). People as sensors: Imputing maps from human actions. In *2018 IEEE/RSJ International Conference on Intelligent Robots and Systems (IROS)* (pp. 2342–2348). IEEE.

Akerly, J. (2015, August). Embodied flow in experiential media systems: A study of the dancer's lived experience in a responsive audio system. In *Proceedings of the 2nd International Workshop on Movement and Computing* (pp. 9–16). ACM.

Apostolos, M. K. (1991). Robot choreography: The paradox of robot motion. *Leonardo, 24*(5), 549–552.

Ashenfelter, K. T., Boker, S. M., Waddell, J. R., & Vitanov, N. (2009). Spatiotemporal symmetry and multifractal structure of head movements during dyadic conversation. *Journal of Experimental Psychology: Human Perception and Performance, 35*(4), 1072–1091.

Atzori, P., & Woolford, K. (1997). Extended-Body: Interview with Stelarc. In *Digital Delirium* (pp. 194–199). Palgrave Macmillan.

Aviezer, H., Trope, Y., & Todorov, A. (2012, November 30). Body cues, not facial expressions, discriminate between intense positive and negative emotions. *Science, 338*(6111), 1225–1229.

Bacula, A., & Knight, H. (2019). Excluded by the JellyFish: Robot-group expressive motion. In *Robotics and Art. International Conference on Robotics and Automation (ICRA)*. https://roboticart.org/wp-content/uploads/2019/05/10_ICRAX_Extended _Abstract_Bacula_Knight_revision.pdf.

Bacula, A., & LaViers, A. (2021). Character synthesis of ballet archetypes on robots using Laban Movement Analysis: Comparison between a humanoid and an aerial robot platform with lay and expert observation. *International Journal of Social Robotics*, *13*(5), 1047–1062.

Badler, N., Costa, M., Zhao, L., & Chi, D. (2000, June). To gesture or not to gesture: What is the question? In *Proceedings Computer Graphics International 2000* (pp. 3–9). IEEE.

Banes, S. (2011). *Terpsichore in Sneakers: Post-modern Dance*. Wesleyan University Press.

Barakova, E. I., & Lourens, T. (2010). Expressing and interpreting emotional movements in social games with robots. *Personal and Ubiquitous Computing*, *14*(5), 457–467.

Barakova, E. I., van Berkel, R., Hiah, L., Teh, Y. F., & Werts, C. (2015, December). Observation scheme for interaction with embodied intelligent agents based on Laban notation. In *2015 IEEE International Conference on Robotics and Biomimetics (ROBIO)* (pp. 2525–2530). IEEE.

Bartenieff, I. (with Lewis, D.). (2013). *Body Movement: Coping with the Environment*. Routledge. (Originally published in 1980.)

Bartneck, C., Belpaeme, T., Eyssel, F., Kanda, T., Keijsers, M., & Šabanović, S. (2020). *Human-Robot Interaction: An Introduction*. Cambridge University Press.

Belta, C., Bicchi, A., Egerstedt, M., Frazzoli, E., Klavins, E., & Pappas, G. J. (2007). Symbolic planning and control of robot motion. *IEEE Robotics & Automation Magazine*, *14*(1), 61–70.

Berl, E., Pakrasi, I., & LaViers, A. (2019, November). Creating context through performance: Perception of the "Dancing Droid" robotic platform in variable valence interactions in distinct office environments. In *International Conference on Social Robotics* (pp. 288–298). Springer.

Bernstein, N. (1967). *The Co-ordination and Regulation of Movements*. Pergamon Press. (Originally published in 1926.)

Berrueta, T. A., Pervan, A., Fitzsimons, K., & Murphey, T. D. (2018). Dynamical system segmentation for information measures in motion. *IEEE Robotics and Automation Letters*, *4*(1), 169–176.

Bishko, L. (2014). Animation principles and Laban movement analysis: Movement frameworks for creating empathic character performances. In T. J. Tanenbaum, M. S. El-Nasr, & M. Nixon (Eds.), *Nonverbal Communication in Virtual Worlds: Understanding and Designing Expressive Characters* (pp. 177–203). ETC Press.

Bloom, K., Adrian, B., Casciero, T., Mizenko, J., & Porter, C. (2017). *The Laban Workbook for Actors: A Practical Training Guide with Video*. Bloomsbury Publishing.

Bradley, K. K. (2008). *Rudolf Laban*. Routledge.

Bregler, C. (1997, June). Learning and recognizing human dynamics in video sequences. In *Proceedings of IEEE Computer Society Conference on Computer Vision and Pattern Recognition* (pp. 568–574). IEEE.

Burton, S. J., Samadani, A. A., Gorbet, R., & Kulić, D. (2016). Laban Movement Analysis and affective movement generation for robots and other near-living creatures. In J. P. Laumond & N. Abe (Eds.), *Dance Notations and Robot Motion* (pp. 25–48). Springer.

Bushman, A., Asselmeier, M., Won, J., & LaViers, A. (2020, May). Toward human-like teleoperated robot motion: Performance and perception of a choreography-inspired method in static and dynamic tasks for rapid pose selection of articulated robots. In *2020 IEEE International Conference on Robotics and Automation (ICRA)* (pp. 10219–10225). IEEE.

Butterworth, J., & Wildschut, L. (Eds.). (2009). *Contemporary Choreography: A Critical Reader*. Routledge.

Calvert, T. W., Bruderlin, A., Mah, S., Schiphorst, T., & Welman, C. (1993, May). The evolution of an interface for choreographers. In *Proceedings of the INTERACT'93 and CHI'93 Conference on Human Factors in Computing Systems* (pp. 115–122). ACM.

Cancienne, M. B., & Snowber, C. N. (2003). Writing rhythm: Movement as method. *Qualitative Inquiry*, *9*(2), 237–253.

Čapek, K. (1923). *RUR (Rossum's Universal Robots): A Fantastic Melodrama*. Doubleday, Page.

Cenciarini, M., & Dollar, A. M. (2011, June). Biomechanical considerations in the design of lower limb exoskeletons. In *2011 IEEE International Conference on Rehabilitation Robotics* (pp. 1–6). IEEE.

Cha, E., Fitter, N. T., Kim, Y., Fong, T., & Mataric, M. J. (2018, March). Effects of robot sound on auditory localization in human-robot collaboration. In *2018 13th ACM/IEEE International Conference on Human-Robot Interaction (HRI)* (pp. 434–442). IEEE.

Clarke, T. J., Bradshaw, M. F., Field, D. T., Hampson, S. E., & Rose, D. (2005). The perception of emotion from body movement in point-light displays of interpersonal dialogue. *Perception*, *34*(10), 1171–1180.

Cohen, B. B., Nelson, L., & Smith, N. S. (2012). *Sensing, Feeling, and Action: The Experiential Anatomy of Body-Mind Centering*. Contact Editions. (Originally published in 1993.)

Crosbie, J., Vachalathiti, R., & Smith, R. (1997). Patterns of spinal motion during walking. *Gait & Posture*, *5*(1), 6–12.

Cuan, C. (2021). Dances with robots: Choreographing, correcting, and performing with moving machines. *TDR*, *65*(1), 124–143.

Cuan, C., Pakrasi, I., Berl, E., & LaViers, A. (2018, August). Curtain and time to compile: A demonstration of an experimental testbed for human-robot interaction. In *2018 27th IEEE International Symposium on Robot and Human Interactive Communication (RO-MAN)* (pp. 255–261). IEEE.

Cuan, C., Berl, E., & LaViers, A. (2019a). Measuring human perceptions of expressivity in natural and artificial systems through the live performance piece *Time to Compile*. *Paladyn, Journal of Behavioral Robotics, 10*(1), 364–379.

Cuan, C., Berl, E., & LaViers, A. (2019b). *Time to Compile:* A performance installation as human-robot interaction study examining self-evaluation and perceived control. *Paladyn, Journal of Behavioral Robotics, 10*(1), 267–285.

Cuan, C., Hoffswell, J., & LaViers, A. (2020, July). Stories about the future: Initial results exploring how co-movement with robots affects perceptions about robot capability. In *Proceedings of the 7th International Conference on Movement and Computing* (Article 10). ACM.

Cui, H., Maguire, C., & LaViers, A. (2019). Laban-inspired task-constrained variable motion generation on expressive aerial robots. *Robotics, 8*(2), 24.

Dahl, L., Bellona, J., Bai, L., & LaViers, A. (2017, June). Data-driven design of sound for enhancing the perception of expressive robotic movement. In *Proceedings of the 4th international conference on movement and computing* (Article 16). ACM.

Darling, K., Nandy, P., & Breazeal, C. (2015, August). Empathic concern and the effect of stories in human-robot interaction. In *2015 24th IEEE International Symposium on Robot and Human Interactive Communication (RO-MAN)* (pp. 770–775). IEEE.

Davis, M. (1983). An introduction to the Davis nonverbal communication analysis system (DaNCAS). *American Journal of Dance Therapy, 6*(1), 49–73.

Delahunta, S., & Shaw, N. Z. (2006). Constructing memories: Creation of the choreographic resource. *Performance Research, 11*(4), 53–62.

Del Vecchio, D., Murray, R. M., & Perona, P. (2003). Decomposition of human motion into dynamics-based primitives with application to drawing tasks. *Automatica, 39*(12), 2085–2098.

Derrida, J. (1974). *Of Grammatology*. JHU Press.

Dickerman, L., & Affron, M. (2012). *Inventing Abstraction, 1910–1925: How a Radical Idea Changed Modern Art*. Museum of Modern Art.

Dils, A. (2002). The ghost in the machine: Merce Cunningham and Bill T. Jones. *PAJ, 24*(1), 94–104.

Dörr, E. (2007). *Rudolf Laban: The Dancer of the Crystal*. Scarecrow Press.

Dourish, P. (2001). *Where the Action Is: The Foundations of Embodied Interaction*. MIT Press.

Dragan, A. D., Lee, K. C., & Srinivasa, S. S. (2013, March). Legibility and predictability of robot motion. In *2013 8th ACM/IEEE International Conference on Human-Robot Interaction (HRI)* (pp. 301–308). IEEE.

Durupinar, F., Kapadia, M., Deutsch, S., Neff, M., & Badler, N. I. (2016). Perform: Perceptual approach for adding ocean personality to human motion using Laban Movement Analysis. *ACM Transactions on Graphics, 36*(1), 1–16.

Eddy, M. (2016). *Mindful Movement: The Evolution of the Somatic Arts and Conscious Action.* Intellect Books.

Elgin, C. (2010). Exemplification and the dance. In J. Beauquel & R. Pouivet (Eds.), *Philosophie de la Danse* (pp. 81–98). Rennes, France: Presses Universitaire de Rennes.

Elswit, K. (2020, July 21). Dancing with our coronasphere to navigate the pandemic. *Dance Magazine.* https://www.dancemagazine.com/six-feet-distance/.

Elswit, K. (2021). Reflections on bodies in lockdown: The coronasphere. *Multimodality & Society, 1*(1), 69–74.

Fagan, R., Conitz, J., & Kunibe, E. (1997). Observing behavioral qualities. *International Journal of Comparative Psychology, 10*(4), 167–179.

Fallatah, A., Urann, J., & Knight, H. (2019, November). The robot show must go on: Effective responses to robot failures. In *2019 IEEE/RSJ International Conference on Intelligent Robots and Systems (IROS)* (pp. 325–332). IEEE.

Fanti, C. (2008). *Towards Automatic Discovery of Human Movemes.* Unpublished doctoral dissertation. California Institute of Technology.

Fdili Alaoui, S., Carlson, K., & Schiphorst, T. (2014, June). Choreography as mediated through compositional tools for movement: Constructing a historical perspective. In *Proceedings of the 2014 international workshop on movement and computing* (pp. 1–6). ACM.

Fdili Alaoui, S., Carlson, K., Cuykendall, S., Bradley, K., Studd, K., & Schiphorst, T. (2015a, August). How do experts observe movement? In *Proceedings of the 2nd international workshop on movement and computing* (pp. 84–91). ACM.

Fdili Alaoui, S., Schiphorst, T., Cuykendall, S., Carlson, K., Studd, K., & Bradley, K. (2015b, June). Strategies for embodied design: The value and challenges of observing movement. In *Proceedings of the 2015 ACM SIGCHI conference on creativity and cognition* (pp. 121–130). ACM.

Fdili Alaoui, S., Françoise, J., Schiphorst, T., Studd, K., & Bevilacqua, F. (2017, May). Seeing, sensing and recognizing Laban movement qualities. In *Proceedings of the 2017 CHI conference on human factors in computing systems* (pp. 4009–4020). ACM.

Fernandes, C. (2014). *The Moving Researcher: Laban/Bartenieff Movement Analysis in Performing Arts Education and Creative Arts Therapies.* Jessica Kingsley Publishers.

Fink, J., Mubin, O., Kaplan, F., & Dillenbourg, P. (2012, May). Anthropomorphic language in online forums about Roomba, AIBO and the iPad. In *2012 IEEE Workshop on Advanced Robotics and Its Social Impacts (ARSO)* (pp. 54–59). IEEE.

Flash, T., & Hogan, N. (1985). The coordination of arm movements: An experimentally confirmed mathematical model. *Journal of Neuroscience*, 5(7), 1688–1703.

Foster, S. L. (1986). *Reading Dancing: Bodies and Subjects in Contemporary American Dance*. University of California Press.

Foster, S. L. (2002). *Dances That Describe Themselves: The Improvised Choreography of Richard Bull*. Wesleyan University Press.

Foster, S. L. (2010). *Choreographing Empathy: Kinesthesia in Performance*. Routledge.

Françoise, J., Fdili Alaoui, S., Schiphorst, T., & Bevilacqua, F. (2014, June). Vocalizing dance movement for interactive sonification of Laban effort factors. In *Proceedings of the 2014 Conference on Designing Interactive Systems* (pp. 1079–1082). ACM.

Frederiksen, M. R., & Stoy, K. (2019, November). A systematic comparison of affective robot expression modalities. In *2019 IEEE/RSJ International Conference on Intelligent Robots and Systems (IROS)* (pp. 1385–1392). IEEE.

Gemeinboeck, P. (2021, March 16). The aesthetics of encounter: A relational-performative design approach to human-robot interaction. *Frontiers in Robotics and AI*, 7, Article 577900.

Gemeinboeck, P., & Saunders, R. (2017, June). Movement matters: How a robot becomes body. In *Proceedings of the 4th International Conference on Movement Computing* (Article 8). ACM.

Geva, N., Uzefovsky, F., & Levy-Tzedek, S. (2020). Touching the social robot PARO reduces pain perception and salivary oxytocin levels. *Scientific Reports*, 10(1), 1–15.

Gielniak, M. J., Liu, C. K., & Thomaz, A. L. (2011, May). Task-aware variations in robot motion. In *2011 IEEE International Conference on Robotics and Automation* (pp. 3921–3927). IEEE.

Gilbreth, F. B., & Gilbreth, L. M. (1919). *Applied Motion Study: A Collection of Papers on the Efficient Method to Industrial Preparedness*. Macmillan.

Gillespie, A., & Zittoun, T. (2013). Meaning making in motion: Bodies and minds moving through institutional and semiotic structures. *Culture & Psychology*, 19(4), 518–532.

Gillies, M. (2009). Learning finite-state machine controllers from motion capture data. *IEEE Transactions on Computational Intelligence and AI in Games*, 1(1), 63–72.

Gladwell, M. (2019). *Talking to strangers: What we should know about the people we don't know*. Penguin UK.

Gladwell, M. (2009). *What the dog saw: And other adventures*. Hachette UK.

Goodman, N. (1976). *Languages of art: An approach to a theory of symbols*. Hackett.

Gray, V. (2015). The choreography of anticipation in Maria Hassabi's PREMIERE. *TDR, 59*(3), 150–157.

Guest, A. H. (2013). *Labanotation: The System of Analyzing and Recording Movement.* Routledge.

Guest, A. H. (2000). *Motif at a Glance! A Quick Guide to Motif Description, the Method of Recording Movement Concepts.* Language of Dance Centre.

Guest, A. H., & Jeschke, C. (1991). *Nijinsky's* Faune *Restored: A Study of Vaslav Nijinsky's 1915 Dance Score* L'Après-midi d'un Faune *and His Dance Notation System: Revealed, Translated into Labanotation and Annotated.* Psychology Press.

Hackney, P. (2003). *Making Connections: Total Body Integration through Bartenieff Fundamentals.* Routledge.

Haffner, N., Kuchelmeister, V., & Ziegler, C. (2012). *William Forsythe: Improvisation Technologies: A Tool for the Analytical Dance Eye* [Film, DVD]. Hatje Cantz. (Originally released in 1999.)

Halprin, A. (1975). Rituals of space: Investing in place. *Journal of Architectural Education, 29*(1), 26–27.

Halprin, L. (2014). The rSVp cycles: Creative processes in the human environment. *Choreographic Practices, 5*(1), 39–47.

Hamburg, J. (1995). Coaching athletes using Laban Movement Analysis. *Journal of Physical Education, Recreation & Dance, 66*(2), 34–37.

Hamburg, J. (Dir.). (2004). *Motivating Moves for People with Parkinson's* [Film, DVD]. Parkinson's Disease Foundation.

Hanson, D. (2005, December). Expanding the aesthetic possibilities for humanoid robots. In *IEEE-RAS International Conference on Humanoid Robots* (pp. 24–31). Citeseer.

Heider, F., & Simmel, M. (1944). An experimental study of apparent behavior. *American Journal of Psychology, 57*(2), 243–259.

Heimerdinger, M., & LaViers, A. (2019). Modeling the interactions of context and style on affect in motion perception: Stylized gaits across multiple environmental contexts. *International Journal of Social Robotics, 11*(3), 495–513.

Hendren, S. (2020). *What Can a Body Do? How We Meet the Built World.* Penguin.

Herath, D. C., Jochum, E., & Vlachos, E. (2017). An experimental study of embodied interaction and human perception of social presence for interactive robots in public settings. *IEEE Transactions on Cognitive and Developmental Systems, 10*(4), 1096–1105.

Herath, D., Kroos, C., & Stelarc (Eds.). (2016). *Robots and Art: Exploring an Unlikely Symbiosis.* Springer Singapore.

Hidalgo, C. (2015). *Why Information Grows: The Evolution of Order, from Atoms to Economies*. Basic Books.

Hirsch, A. B. (2011). Scoring the participatory city: Lawrence (& Anna) Halprin's Take Part Process. *Journal of Architectural Education, 64*(2), 127–140.

Höök, K. (2018). *Designing with the Body: Somaesthetic Interaction Design*. MIT Press.

Höök, K., Caramiaux, B., Erkut, C., Forlizzi, J., Hajinejad, N., Haller, M., Hummels C., Isbister K., Jonsson, M., Khut, G., Loke, L., & Tobiasson, H. (2018, March). Embracing first-person perspectives in soma-based design. *Informatics, 5*(1), 8.

Höök, K., Benford, S., Tennent, P., Tsaknaki, V., Alfaras, M., Avila, J. M., Li, C., Marshall, J., Daudén Roquet, C., Sanches, P., Ståhl, A., Umair, M., Windlin, C., & Zhou, F. (2021). Unpacking non-dualistic design: The soma design case. *ACM Transactions on Computer-Human Interaction, 28*(6), 1–36.

Humphrey, D. & Pollack, B. (1956). *The Art of Making Dances*. Rinehart.

Huzaifa, U., Bernier, C., Calhoun, Z., Heddy, G., Kohout, C., Libowitz, B., Moenning, A., Ye, J., Maguire, C., & LaViers, A. (2016, June). Embodied movement strategies for development of a core-located actuation walker. In *2016 6th IEEE International Conference on Biomedical Robotics and Biomechatronics (BioRob)* (pp. 176–181). IEEE.

Huzaifa, U., Fuller, C., Schultz, J., & LaViers, A. (2019, November). Toward a bipedal robot with variable gait styles: Sagittal forces analysis in a planar simulation and a prototype ball-tray mechanism. In *2019 IEEE/RSJ International Conference on Intelligent Robots and Systems (IROS)* (pp. 2266–2272). IEEE.

Huzaifa, U., Maguire, C., & LaViers, A. (2020). Toward an expressive bipedal robot: Variable gait synthesis and validation in a planar model. *International Journal of Social Robotics, 12*(1), 129–141.

Isaacson, W. (2014). *The Innovators: How a Group of Inventors, Hackers, Geniuses and Geeks Created the Digital Revolution*. Simon and Schuster.

Isaacson, W. (2011). *Steve Jobs*. Simon and Schuster.

Jang Sher, A. (2017). An embodied, platform-invariant architecture for robotic spatial commands. [Unpublished master's thesis]. University of Illinois Urbana-Champaign.

Jang Sher, A., Huzaifa, U., Li, J., Jain, V., Zurawski, A., & LaViers, A. (2019). An embodied, platform-invariant architecture for connecting high-level spatial commands to platform articulation. *Robotics and Autonomous Systems, 119*, 263–277.

Jeon, M., Fiebrink, R., Edmonds, E. A., & Herath, D. (2019). From rituals to magic: Interactive art and HCI of the past, present, and future. *International Journal of Human-Computer Studies, 131*, 108–119.

Jochum, E., Vlachos, E., Christoffersen, A., Nielsen, S. G., Hameed, I. A., & Tan, Z. H. (2016). Using theatre to study interaction with care robots. *International Journal of Social Robotics, 8*(4), 457–470.

Jochum, E., & Goldberg, K. (2016). Cultivating the uncanny: The Telegarden and other oddities. In D. Herath, C. Kroos, & Stelarc (Eds.), *Robots and Art: Exploring an Unlikely Symbiosis* (pp. 149–175). Springer Singapore.

Joo, H., Liu, H., Tan, L., Gui, L., Nabbe, B., Matthews, I., Kanade, T., Nobuhara, S., & Sheikh, Y. (2015). Panoptic studio: A massively multiview system for social motion capture. In Proceedings of the *IEEE International Conference on Computer Vision* (pp. 3334–3342). IEEE.

Kaushik, R., & LaViers, A. (2019). Imitation of human motion by low degree-of-freedom simulated robots and human preference for mappings driven by spinal, arm, and leg activity. *International Journal of Social Robotics, 11*(5), 765–782.

Kaushik, R., Vidrin, I., & LaViers, A. (2018). Quantifying coordination in human dyads via a measure of verticality. In *Proceedings of the 5th International Conference on Movement and Computing* (Article 19). ACM.

Kelly, T. F. (2014). *Capturing Music: The Story of Notation.* W. W. Norton.

Kim, H. J., Neff, M., & Lee, S. H. (2022). The perceptual consistency and association of the LMA effort elements. *ACM Transactions on Applied Perception, 19*(1), 1–17.

Kim, J., Seo, J. H., & Kwon, D. S. (2012, November). Application of effort parameter to robot gesture motion. In *2012 9th International Conference on Ubiquitous Robots and Ambient Intelligence (URAI)* (pp. 80–82). IEEE.

King, K. (2020, February 21). The robot in aisle five isn't stalking you. No, really. *Wall Street Journal.* https://www.wsj.com/articles/the-robot-in-aisle-five-isnt-stalking -you-no-really-11582302075.

Kingston, P., & Egerstedt, M. (2011, June). Motion preference learning. In *Proceedings of the 2011 American Control Conference* (pp. 3819–3824). IEEE.

Knight, H., & Simmons, R. (2014, August). Expressive motion with x, y and theta: Laban effort features for mobile robots. In *23rd IEEE International Symposium on Robot and Human Interactive Communication* (pp. 267–273). IEEE.

Kozel, S. (2007). *Closer: Performance, Technologies, Phenomenology.* MIT Press.

Laban, R. (1966). *Choreutics.* London: Macdonald & Evans.

Laban, R., & Lawrence, F. C. (1959). *Effort.* Macdonald & Evans.

Laban, R., and Ullmann, L. (2011). *The Mastery of Movement.* Dance Books. (Originally published in 1950.)

Ladenheim, K., & LaViers, A. (2021, March 18). Babyface: Performance and installation art exploring the feminine ideal in gendered machines. *Frontiers in Robotics and AI*, 8, Article 576664.

Lagriffoul, F., Dantam, N. T., Garrett, C., Akbari, A., Srivastava, S., & Kavraki, L. E. (2018). Platform-independent benchmarks for task and motion planning. *IEEE Robotics and Automation Letters*, *3*(4), 3765–3772.

Lamb, W., & Watson, E.M. (1979). *Body Code: The Meaning in Movement.* Taylor & Francis.

Lambert, J., Huzaifa, U., Rizvi, W., & LaViers, A. (2019, October). A comparison of descriptive and emotive labels to explain human perception of gait styles on a compass walker in variable contexts. In *2019 28th IEEE International Conference on Robot and Human Interactive Communication (RO-MAN)* (pp. 1–8). IEEE.

Laumond, J. P., & Abe, N. (Eds.). (2016). *Dance Notations and Robot Motion.* Springer International.

Laurier, E. (2013). Capturing motion: Video set-ups for driving, cycling and walking. In *Routledge Handbook of Mobilities*, (pp. 493–502). Routledge.

LaViers, A. (2013). *Choreographic Abstractions for Style-Based Robotic Motion.* Unpublished doctoral dissertation. Georgia Institute of Technology.

LaViers, A. (2019a, May 8). Counts of mechanical, external configurations compared to computational, internal configurations in natural and artificial systems. *PLoS One*, *14*(5), Article e0215671.

LaViers, A. (2019b, May 23). Ideal mechanization: Exploring the machine metaphor through theory and performance. *Arts*, *8*(2), 67.

LaViers, A. (2019c, January 21). Make robot motions natural: Humanoid machines should move and gesture more like us, argues Amy LaViers. *Nature*, 565, 422–424.

LaViers, A., Cuan, C., Maguire, C. (2018). Choreographic and somatic approaches for the development of expressive robotic systems. *Arts*, *7*(2), 11.

LaViers, A., & Egerstedt, M. (2011, June). The ballet automaton: A formal model for human motion. In *Proceedings of the 2011 American Control Conference* (pp. 3837–3842). IEEE.

LaViers, A., & Egerstedt, M. (2014, April). Style-based abstractions for human motion classification. In *2014 ACM/IEEE International Conference on Cyber-physical Systems (ICCPS)* (pp. 84–91). IEEE.

LaViers, A., & Egerstedt, M. (2012, June). Style based robotic motion. In *Proceedings of the 2012 American control conference* (pp. 4327–4332). IEEE.

LaViers, A., Teague, L., & Egerstedt, M. (2014). Style-based robotic motion in contemporary dance performance. In A. LaViers & M. Egerstedt (Eds.), *Controls and Art: Inquiries at the Intersection of the Subjective and the Objective* (pp. 205–229). Springer Cham.

LaViers, A., & Maguire, C. (2022). The BESST System: Explicating a new component of time in Laban/Bartenieff Movement Studies through work with robots. In *Proceedings of the 8th International Conference on Movement and Computing* (Article 40). ACM.

Lazier, R., & Trueman, D. (2016). *There Might Be Others.* Operating System Press.

Luo, Y., Ye, J., Adams, R. B., Li, J., Newman, M. G., & Wang, J. Z. (2020). Arbee: Towards automated recognition of bodily expression of emotion in the wild. *International Journal of Computer Vision, 128*(1), 1–25.

Maletic, V. (2011). *Body-Space-Expression: The Development of Rudolf Laban's Movement and Dance Concepts.* Walter de Gruyter.

Majid, A., & Kruspe, N. (2018). Hunter-gatherer olfaction is special. *Current Biology, 28*(3), 409–413.

Majid, A., Burenhult, N., Stensmyr, M., De Valk, J., & Hansson, B. S. (2018). Olfactory language and abstraction across cultures. *Philosophical Transactions of the Royal Society B: Biological Sciences, 373*(1752), Article 20170139.

Männistö-Funk, T., & Sihvonen, T. (2018). Voices from the uncanny valley. *Digital Culture & Society, 4*(1), 45–64.

Marrocco, W. T. (1964). The notation in American sacred music collections. *Acta Musicologica, 36*(2/3), 136–142.

McRae, C., Leventhal, D., Westheimer, O., Mastin, T., Utley, J., and Russell, D. (2018). Long-term effects of Dance for PD® on self-efficacy among persons with Parkinson's disease. *Arts & Health, 10*(1), 85–96.

Merleau-Ponty, M. (1945). *Phenomenology of Perception.* Motilal Banarsidass.

Michalowski, M. P., Šabanović, S., & Kozima, H. (2007, March). A dancing robot for rhythmic social interaction. In *Proceedings of the ACM/IEEE International Conference on Human-Robot Interaction* (pp. 89–96). ACM.

Michalowski, M. P., Simmons, R., & Kozima, H. (2009, September). Rhythmic attention in child-robot dance play. In *the 18th IEEE International Symposium on Robot and Human Interactive Communication* (pp. 816–821). IEEE.

Milzoff, R. (2020, November 7). Inside "Single Ladies" choreographer JaQuel Knight's quest to copyright his dances. *Billboard.* https://www.billboard.com/music/music

-news/jaquel-knight-beyonce-megan-thee-stallion-billboard-cover-story-interview
-2020-9477613/.

Mims, Christopher. (2021, June 5). Self-driving cars could be decades away, no matter what Elon Musk said. *Wall Street Journal*. https://www.wsj.com/articles/self-driving-cars-could-be-decades-away-no-matter-what-elon-musk-said-11622865615.

Mori, M., MacDorman, K. F., & Kageki, N. (2012). The uncanny valley. *IEEE Robotics & Automation Magazine, 19*(2), 98–100.

Muybridge, E. (1887). *Animal Locomotion*. Da Capo.

Nakaoka, S., Nakazawa, A., Yokoi, K., & Ikeuchi, K. (2004, April). Leg motion primitives for a dancing humanoid robot. In *IEEE International Conference on Robots and Automation, 2004. Proceedings. ICRA'04. 2004* (Vol. 1, pp. 610–615). IEEE.

Nakata, T., Sato, T., Mori, T., & Mizoguchi, H. (1998, June). Expression of emotion and intention by robot body movement. In *International Conference on Intelligent Autonomous Systems 5 (IAS-5)* (pp. 352–359). IOS Press.

Nakata, T. (2001, September). Algorithmic choreography of animal-like body movement based on Bartenieff's and Kestenberg's theories. In *Proceedings of the 10th IEEE International Workshop on Robot and Human Interactive Communication* (pp. 207–212). IEEE.

Nettl-Fiol, R., & Vanier, L. (2011). *Dance and the Alexander Technique: Exploring the Missing Link*. University of Illinois Press.

Newlove, J. (2007). *Laban for Actors and Dancers*. Nick Hern Books.

Newlove, J., & Dalby, J. (2004). *Laban for All*. Routledge.

Niebles, J. C., & Fei-Fei, L. (2007, June). A hierarchical model of shape and appearance for human action classification. In *2007 IEEE Conference on Computer Vision and Pattern Recognition* (pp. 1–8). IEEE.

Nin, A. (1961). *Seduction of the Minotaur*. Swallow Press.

Notomista, G., Emam, Y., & Egerstedt, M. (2019). The SlothBot: A novel design for a wire-traversing robot. *IEEE Robotics and Automation Letters, 4*(2), 1993–1998.

Ofli, F., Chaudhry, R., Kurillo, G., Vidal, R. & Bajcsy, R. (2014). Sequence of the Most Informative Joints (SMIJ): A new representation for human skeletal action recognition. *Journal of Visual Communication and Image Representation, 25*(1), 24–38.

Olsen, A., & McHose, C. (2004). *Bodystories: A Guide to Experiential Anatomy*. UPNE.

Özcimder, K., Dey, B., Lazier, R. J., Trueman, D., & Leonard, N. E. (2016, July). Investigating group behavior in dance: An evolutionary dynamics approach. In *Proceedings of the 2016 American Control Conference* (pp. 6465–6470). IEEE.

Özcimder, K., Dey, B., Franci, A., Lazier, R., Trueman, D., & Leonard, N. E. (2019). Social decision-making driven by artistic explore–exploit tension. *Interdisciplinary Science Reviews, 44*(1), 55–81.

Pakrasi, I., Chakraborty, N., Cuan, C., Berl, E., Rizvi, W., & LaViers, A. (2018, November). Dancing droids: An expressive layer for mobile robots developed within choreographic practice. In *International Conference on Social Robotics* (pp. 410–420). Springer Cham.

Palazzi, M., Shaw, N. Z., Forsythe, W., Lewis, M., Albright, B., Andereck, M., Bhatawadekar, S., Ban, H., Calhoun, A., Drozd, J., Fry, J., Quintanilha, M., Reed, A., Schroeder, B., Skove, L., Thorndike, A., Twohig, M., Ahlqvist, O., Chan, P., Cressie, N., Turk, S., Johnson, J., Roman, C., Waterhouse, E., deLahunta, S., Haggard, P., & Noe, A. (2009). Synchronous objects for one flat thing, reproduced. In *ACM SIGGRAPH 2009 Art Gallery*, 37. ACM.

Park, E., & Lee, J. (2014). I am a warm robot: The effects of temperature in physical human–robot interaction. *Robotica, 32*(1), 133–142.

Peng, X. B., Kanazawa, A., Malik, J., Abbeel, P., & Levine, S. (2018). SFV: Reinforcement learning of physical skills from videos. *ACM Transactions on Graphics, 37*(6), 1–14.

Raindel, N., Liron, Y., & Alon, U. (2021, June 7). A study of dramatic action and emotion using a systematic scan of stick figure configurations. *Frontiers in Physics, 9,* Article 664948.

Raji, I. D., Gebru, T., Mitchell, M., Buolamwini, J., Lee, J., & Denton, E. (2020, February). Saving face: Investigating the ethical concerns of facial recognition auditing. In *Proceedings of the AAAI/ACM Conference on AI, Ethics, and Society* (pp. 145–151). ACM.

Rajko, J. (2021). *ACM DL Corpus Dataset: A Systematic Mapping Study on Computing Research Involving Dance.* Harvard Dataverse, V1. https://doi.org/10.7910/DVN/RPURHV.

Reeves, B., & Nass, C. (1996). *The Media Equation: How People Treat Computers, Television, and New Media Like Real People.* Cambridge University Press.

Rett, J., & Dias, J. (2007, June). Human-robot interface with anticipatory characteristics based on Laban Movement Analysis and Bayesian models. In *2007 IEEE 10th International Conference on Rehabilitation Robotics* (pp. 257–268). IEEE.

Russell, J. A. (1980). A circumplex model of affect. *Journal of Personality and Social Psychology, 39*(6), 1161.

Šabanović, S., Bennett, C. C., Chang, W. L., & Huber, L. (2013, June). PARO robot affects diverse interaction modalities in group sensory therapy for older adults with dementia. In *2013 IEEE 13th International Conference on Rehabilitation Robotics (ICORR)* (pp. 1–6). IEEE.

Saenko, K., Packer, B., Chen, C., Bandla, S., Lee, Y., Jia, Y., Niebles, J., Koller, D., Fei-Fei, L., Grauman, K., & Darrell, T. (2012, November 14). *Mid-level Features Improve Recognition of Interactive Activities*. Technical report UCB/EECS-2012–209. Department of Electrical Engineering and Computer Sciences, University of California Berkeley.

Salaris, P., Abe, N., & Laumond, J. P. (2017). Robot choreography: The use of the kinetography Laban system to notate robot action and motion. *IEEE Robotics & Automation Magazine, 24*(3), 30–40.

Samaritter, R., & Payne, H. (2017). Through the kinesthetic lens: Observation of social attunement in autism spectrum disorders. *Behavioral Sciences, 7*(1), 14.

Santos, L., Prado, J. A., & Dias, J. (2009, October). Human robot interaction studies on Laban human movement analysis and dynamic background segmentation. In *2009 IEEE/RSJ International Conference on Intelligent Robots and Systems* (pp. 4984–4989). IEEE.

Schøyen Collection. (2022, December 15). *MS 5105*. The Schøyen Collection. https://www.schoyencollection.com/music-notation/old-babylonia-cuneiform-notation/oldest-known-music-notation-ms-5105

Shafir, T., Tsachor, R. P., & Welch, K. B. (2016, January 11). Emotion regulation through movement: Unique sets of movement characteristics are associated with and enhance basic emotions. *Frontiers in Psychology, 6*, Article 2030.

Shannon, C. (1948). A mathematical theory of communication. *Bell System Technical Journal, 27*(3), 379–423.

Sharma, M., Hildebrandt, D., Newman, G., Young, J. E., & Eskicioglu, R. (2013, March). Communicating affect via flight path exploring use of the Laban effort system for designing affective locomotion paths. In *2013 8th ACM/IEEE International Conference on Human-Robot Interaction (HRI)* (pp. 293–300). IEEE.

Shaw, N. Z. (2016). Synchronous objects. In M. Bleeker (Ed.), *Transmission in Motion: The Technologizing of Dance* (pp. 99–107). Routledge.

Sheets-Johnstone, M. (2011). *The Primacy of Movement* (Vol. 2). John Benjamins Publishing.

Sheng, Y., & LaViers, A. (2014, October). Style-based human motion segmentation. In *2014 IEEE International Conference on Systems, Man, and Cybernetics (SMC)* (pp. 240–245). IEEE.

Sicchio, K. (2014). Hacking choreography: Dance and live coding. *Computer Music Journal, 38(1)*, 31–39.

Simmons, R., & Knight, H. (2017). Keep on dancing: Effects of expressive motion mimicry. In *2017 26th IEEE International Symposium on Robot and Human Interactive Communication (RO-MAN)* (pp. 720–727). IEEE.

Singh, A., & Young, J. E. (2012, March). Animal-inspired human-robot interaction: A robotic tail for communicating state. In *2012 7th ACM/IEEE International Conference on Human-Robot Interaction (HRI)* (pp. 237–238). IEEE.

Skybetter, S. (2016, June 10). *The Choreography of the Internet of Things* [Video]. TEDx-Providence. https://www.youtube.com/watch?v=zoNA1dP8_rI.

Skybetter, S. (2020, September 23). Meet the choreographic interface designer who brings her knowledge to Google. *Dance Magazine*. https://www.dancemagazine.com /interaction-design/.

Stark, L., & Crawford, K. (2019). The work of art in the age of artificial intelligence: What artists can teach us about the ethics of data practice. *Surveillance & Society, 17*(3/4), 442–455.

Stedge, C. (2017). *LMA Certification Thesis on Breath*. Unpublished thesis. Laban/ Bartenieff Institute of Movement Studies.

Studd, K. & Cox, L. (2020). *EveryBody Is a Body*. 2nd ed. Outskirts Press. (Originally published in 2013.)

Sutil, N. S. (2015). *Motion and Representation: The Language of Human Movement*. MIT Press.

Swaminathan, D., Thornburg, H., Mumford, J., Rajko, S., James, J., Ingalls, T., Campana, E., Qian, G., Sampath, P., & Peng, B. (2009). A dynamic Bayesian approach to computational Laban shape quality analysis. *Advances in Human-Computer Interaction,* 2009, Article 362651.

Thomas, F., Johnston, O., & Thomas, F. (1995). *The Illusion of Life: Disney Animation*. Hyperion.

Tsachor, R. P., & Shafir, T. (2019, March 28). How shall I count the ways? A method for quantifying the qualitative aspects of unscripted movement with Laban movement analysis. *Frontiers in Psychology, 10*, Article 572.

Vanrie, J., & Verfaillie, K. (2004). Perception of biological motion: A stimulus set of human point-light actions. *Behavior Research Methods, Instruments, & Computers, 36*(4), 625–629.

Varella, M. A. C. (2018, October 1). The biology and evolution of the three psychological tendencies to anthropomorphize biology and evolution. *Frontiers in Psychology, 9*, Article 1839.

Wahl, C. (2019). *Laban/Bartenieff Movement Studies: Contemporary Applications*. Human Kinetics.

Wang, P., Li, W., Ogunbona, P., Wan, J., & Escalera, S. (2018). RGB-D-based human motion recognition with deep learning: A survey. *Computer Vision and Image Understanding, 171*, 118–139.

Werner, H., & Kaplan, B. (1984). *Symbol Formation*. Psychology Press.

Whittier, C. (2017). *Creative Ballet Teaching: Technique and Artistry for the 21st Century Ballet Dancer*. Routledge.

Williams, D. (2007). On choreometrics. *Visual Anthropology*, *20*, 233–239.

Wong, M., & Danesi, M. (2015). Color, shape, and sound: A proposed system of music notation. *Semiotica*, *2015*(204), 419–428.

Worth, L., & Poynor, H. (2018). *Anna Halprin*. Routledge.

Young, J. (2017). The therapeutic movement relationship in dance/movement therapy: A phenomenological study. *American Journal of Dance Therapy*, *39*(1), 93–112.

Zhang, X., & Wakkary, R. (2014, June). Understanding the role of designers' personal experiences in interaction design practice. In *Proceedings of the 2014 Conference on Designing Interactive Systems* (pp. 895–904). ACM.

Zhou, A., & Dragan, A. D. (2018, October). Cost functions for robot motion style. *In 2018 IEEE/RSJ International Conference on Intelligent Robots and Systems (IROS)* (pp. 3632–3639). IEEE.

Zhou, Y., Asselmeier, M., & LaViers, A. (2019, October). Toward expressive multi-platform teleoperation: Laban-inspired concurrent operation of multiple joints on the Rethink Robotics Baxter robot in static and dynamic tasks. In *Proceedings of the 6th International Conference on Movement and Computing* (Article 15). ACM.

Ziegelmaier, R. S., Correia, W., Teixeira, J. M., & Simões, F. P. (2020, November). Components of the LMA as a design tool for expressive movement and gesture construction. In *2020 22nd Symposium on Virtual and Augmented Reality* (pp. 321–330). IEEE.

Zordan, V. B., Celly, B., Chiu, B., & DiLorenzo, P. C. (2004, August). Breathe easy: Model and control of simulated respiration for animation. *In Proceedings of the 2004 ACM SIGGRAPH/Eurographics Symposium on Computer Animation* (pp. 29–37). ACM.

Index